Women Drive Tractors Too

18 True Stories of Irish Women in Agriculture

Farm yards are not playgrounds. Always supervise children.

ISBN: 1-905451-03-2

A CIP catalogue for this book is available from the National Library.

Cover Design: Once Upon Design, Drogheda
Layout: Valerie Seery
Printed in Ireland by Johnswood Press Ltd.

This book was published in cooperation with Choice Publishing & Book Services Ltd, Ireland
Tel: 041 9841551 Email: info@choicepublishing.ie
www.choicepublishing.ie

Dedication

To my mother Eileen Byrne Carroll, a true farm woman and heroine, who continues to inspire me every day.

Mary Carroll is former Equality Office with the Irish Farmers Association and is now working as a consultant in the area of communications, training and equality, including a part-time position with Concern.

She produces and presents a weekly evening show on local radio station Tipp FM and manages editorial content for special supplements with a variety of newspapers.

Mary qualified with a BSc in Zoology and a MSc in Environmental Science. She also holds qualifications in legal studies and PR.

Mary continues to live in her home village of Ballyroan, Co. Laois.

Women Drive Tractors Too

18 True Stories of Irish Women in Agriculture

Contents

Women Drive Tractors Too

18 True Stories of Irish Women in Agriculture

By Mary Carroll

Introduction

Close your eyes and conjure up the image of a farmer....it is invariably a man in a field, a man on a tractor....always a man! As an only daughter who grew up on a farm in County Laois in Ireland, I was always very interested in agriculture but always acutely aware of this stereotype. My experiences as a Farm Advisor with the Department of Agriculture and Rural Affairs (DARD) in Northern Ireland, further brought home to me just how invisible women in agriculture can be.

As the IFA Equality Officer, I had the opportunity to work with some outstanding women in agriculture. It was their inspiration, and my belief that women deserve more recognition for their contribution to farming, that has led to this book.

I was associated with a number of significant events during my time as IFA Equality Officer, including the Croke Park Conference 2003, the Tractorcade 2003 and the trip to Australia in 2004. These events appear throughout the stories in the book.

The inaugural IFA conference for 'Women in Agriculture' in Dublin's Croke Park Conference Centre, October 2003, supported by ACCBank, was a landmark day. Almost 600 farm women from all over Ireland attended that memorable event and issued a declaration calling on the Irish Government to establish a 'Women in Agriculture Unit' based on the highly successful Australian model.

In fact links with Australia were very significant. World leaders in terms of positive action for women in agriculture, I turned to them as a role model from the outset of the Equality project. An Australian delegation to the Croke Park conference was followed by a reciprocal visit to Australia by 32 Irish farm women in 2004. Again, many of the women featured in this book made this life changing, inspirational trip to Australia.

The IFA Tractorcade during January 2003 was a pivotal point for many of the women in this book.

I chose these particular 18 role models because they are women who inspire, motivate, engender confidence and promote positivity – not just for me but for everyone with whom they have contact. They come from all over Ireland, from different age groups and from different backgrounds. Some of them married into farms, some of them inherited farms, and some of them bought farms. Some of them have worked full time on their farms, some part time and others have worked off the farm.

Almost all of them had to battle the odds, overcome obstacles and show great determination to get to where they are now. They are dynamic, progressive, assertive and confident women who are willing and able to speak up and to speak out.

When I look at the stories I have gathered here, I can pick out several themes. The absence or presence of strong role models in childhood; frustration at the way men dominate agricultural and other organisations; a decision to become involved in agricultural politics and to change things from the inside; a willingness to broach controversial issues; a hunger for more knowledge; personal tragedies, especially the early loss of a parent; the realisation that not all women in agriculture want women to be more involved in mainstream issues; the challenge for women in a male-dominated environment; the importance of networking; a life-long commitment to change and dealing with adversity.

'Women Drive Tractors Too' highlights some of the age-old issues surrounding a woman's role on the farm. Many of the women in this book came to farm management by default – not through choice or inheritance, and young women today are still less likely to find themselves in a farming career. Despite the positive role models in this book, the image of the farmer continues to be male.

For me, this book is a vehicle for immortalising the triumphs and experiences of women in agriculture. It is an opportunity to hear their voices. These are stories of women who have all made such a difference. Some of them have already been featured in the media, but most of them simply go about their lives quietly with no fuss.

I know that on its own, 'Women Drive Tractors Too', will not make Irish farm women more visible. The cause of making women in agriculture more visible, and a more integral part of the industry, belongs to the Department, the Support

Agencies and agri-political organisations. I sincerely hope they are inspired to rise to the challenge.

In the words of Winston Churchill…

> *"There comes a special moment in everyone's life, a moment for which that person was born. That special opportunity, when they seize it, will fulfil their mission – a mission for which they are uniquely qualified. In that moment they find greatness. It is their finest hour"*

Acknowledgements…

ACCBank is a natural partner in this book and sponsor for its production. The Bank has a long tradition of partnering farmers and the agri-sector. With its roots firmly in farm banking, the Bank continues to expand into the agri business sector. Through its active involvement in some of the Equality Project initiatives, ACCBank has also proved itself to be hugely aware of the potential contribution of women to the farm economy. Special thanks for Paddy Horgan and Hailey Kierse for their encouragement and belief that this book was worth publishing.

My thanks to the LEADER organisations in Laois, Meath and Cavan/Monaghan for their financial support. The aim of LEADER is to encourage and help rural people to think about the longer-term potential of their area.

The IFA Equality Initiative (2002-2004), which was co-funded under the NDP 'Equality for Women' measure, sought to increase participation by women in agricultural decision making. My involvement with the IFA as Equality Officer allowed me to meet these wonderful women, and to capture their stories, and therefore leave a permanent and positive legacy.

An Australian book 'Ordinary People Extraordinary Lives- inspiring stories from rural Australia' was published in 2001 by Margaret Carroll, and proved to be the catalyst that compelled me to write this book and capture these women's stories.

Communication expert and trainer Yanky Fachler helped me edit the material for this book, and Gráinne Harte managed production and marketing, as well as providing ongoing advice and support.

My thanks to Anna Rackard, who took the photos for this book.

To my parents, Johns and Eileen, who, without knowing, really helped inspire me to write the book.

And of course to the 18 women featured, without whom none of this would have been possible.

Foreword

by Mary Coughlan,
TD, Minister for Agriculture and Food

There is no doubt that the women of rural Ireland have been the backbone of farm families down the generations. They often endured considerable hardships and made major sacrifices so that their families were properly cared for and their children educated. Their desire was always to see the next generation having more opportunity for happiness and prosperity than the generation which preceded it. They provided the social infrastructure on which much of the advancement in rural Ireland was made possible.

I come from a farming background, and when I was growing up, my mother was the farming expert in the household, and I always remember her doing the paperwork. However, the nature of farming and life in rural Ireland has changed greatly over recent decades. Despite, or perhaps because of these changes, women continue to play a pivotal role. Whether as business partners in the farm enterprise; carers of children or elderly people; owners, managers or employees in agri-business or the ever-increasing range of service enterprises to be found in the towns and villages of rural Ireland; or within voluntary and other organisations, women continue to make an enormous contribution to sustaining and enhancing life throughout rural Ireland. That contribution encompasses all areas of economic and social activity. The women of rural Ireland are critical in sustaining and improving the good quality of rural life.

The stories that Mary has so compellingly compiled are eloquent testimony to the many roles played by these women, as they make their way through life in a dramatically changing landscape. She has captured the very essence of modern Irish rural life through the eyes of these energetic and enterprising women. But they are also stories of women who have a deep sense of community, and the riches that engagement with local communities can bring. This book provides a fascinating, and deeply personal insight into the women's lives, and the families and communities that shaped them. Through these stories we recognise too our own mothers, sisters, neighbours and friends. In short, their stories are inspiring and in giving us this insight into the lives of just some of the many champions of rural Ireland; Mary has done a service to women everywhere.

Ann Kehoe

I would have to say that Ann must be one of the most amazing women in agriculture that I have ever met. She has a fun personality, and I always have to chuckle when I hear her loving references to her husband: "Brian Kehoe tells me several times a day how much he loves me and how proud he is of me. I am so lucky!"

Ann is someone I have always been able to rely on for strategic planning, and political insights. Whenever I have faced crisis situations in my work, I have always been able to ring Ann, knowing that she will show me a new perspective.

"Perseverance is my middle name," says Ann Kehoe. From the first time I encountered the indomitable South Tipperary farmer known in farming and media circles alike as "the sheep woman," I have tended to agree with this self-assessment. It was dogged determination that helped Ann overcome intense opposition when she decided to so something about the widespread illegal smuggling of lambs. And it was her perseverance that helped bring the issue into the public domain some years ago.

The desire to passionately embrace farming causes was not immediately apparent as Ann grew up as the third child of a South Tipp farming family. "My father had a farming background, but because he had not inherited the family farm, he earned his living working for others. He bought his first farm when I was three years old. When I was six, he bought the farm where I and my siblings were raised."

Ann loved the freedom of the farm, and spent a carefree childhood in a fun, easy-going household. She remembers swimming with friends in the nearby river, and bringing in the hay in the summer. Ann's father loved the sea, and frequently took his family to seaside resorts. He was also passionate about horseracing, and Ann loved the family outings to race meetings around the country.

Ann's idyllic childhood ended when she won a scholarship to a small rural convent boarding school. Her pride in being granted a scholarship was tempered by rebellion against the restrictiveness and drabness of the school. For someone of Ann's spirit, the school was far too quiet.

To brighten things up, Ann and her best friend used to sneak out at night and go to dances. But their decision to notch their rebelliousness up a level and to start smoking, got them into serious trouble. Ann's friend got caught and was expelled. Unwilling to remain in the school without her friend and co-conspirator, with whom she had shared so many adventures and misadventures, Ann decided to use the opportunity to make her own escape, and the two girls ran away together.

They went missing for two whole days, at exactly the same time as a national man-hunt for a kidnapped business executive. When the girls first spotted the massive uniformed presence, they thought the police were looking for them!

Ann remembers that her father adopted a very lax attitude to the whole escapade. "When the two of us finally turned up on the doorstep of my home, my father was very understanding. He had always felt that boarding school was too regimental for youthful spirits. I was then allowed to leave boarding school, and I attended a regular local all-girls secondary school. I found it quite lonely. The other girls were clannish, and I never really established lasting friendships there."

Ann passed her Leaving Certificate, but she still had no clear idea of what direction to take. She had no great ambitions for any particular career, and there was very little guidance available at the time. Choosing the path of least resistance, Ann enrolled for an Advanced Secretarial Course at Waterford Regional College.

After completing her studies, Ann drifted into secretarial work, and was fortunate enough to find a safe and secure job with the local County Council. Luckily, she worked under an enlightened boss who encouraged her to expand her horizons and to take on more and more responsibilities. Ann's work brought her into contact with development agencies like the Industrial Development Agency (IDA), and she got her first taste of the way local bureaucracies operated. The realisation gradually developed that if only she found a suitable avenue to channel her energies, she was probably capable of achieving much more.

It was at this juncture that Brian Kehoe, a friend of Ann's brother, entered the picture. "I still remember the first time Brian drove his silver sports car into our yard. I was immediately struck by this guy whose jet-black curly hair was a little longer than convention demanded. Even before I nurtured any romantic thoughts about him, I had already decided that he was the most exotic creature I had ever set eyes upon."

After gentle nudging by friends ensured that the two kept meeting, Ann reached the inescapable conclusion that she could not imagine spending the rest of her life without Brian. She decided to take the initiative, and in short order they were wed.

Brian had established his own cattle delivery business at the age of 16, while still working on his family farm some 6 miles from Ann's family farm. Ann and Brian built their own home on his father's property, on a perfect spot offering majestic mountain vistas and stunning open spaces. The only fly in the ointment was the nearby pharmaceutical plant that was causing havoc with the health of so many families in the vicinity.

For the first few months after they were married, Ann continued to work at the County Council. "When baby Grace arrived, I was really looking forward to full-time motherhood, and I relished the prospect of several more babies. This was not to be. I had several miscarriages, and I was dogged by ill health. Finally, the consultant at the maternity hospital informed me that it was dangerous and unlikely that I would be able to have another baby."

Once Ann's health improved, the Kehoe threesome of Ann, Brian and Grace enthusiastically embraced the farming life. They travelled round the country, attending farmers meetings and farm open days. Ann and Brian were always interested in what others were doing and how they were doing it. They were eager to learn useful and practical tips, and were interested in everything pertaining to the farming world. Farming was the context that framed their family life, their social life, and their economic life.

It was almost inevitable that eventually this total immersion in the world of farming would bring them to the Irish Farmers Association - IFA. Ann was in her early thirties when friends recommended that she and Brian should attend their local IFA branch meeting. The guest speaker at that first meeting was Henry Britton, who later became one of their closest friends. It was Henry who encouraged them to go along to their first IFA county meeting, at a time when there were precious few women in IFA.

Ann was regarded as something of a novelty, and she was in her element. Here was an opportunity to do something on behalf of a cause that she felt passionately about – sheep. She also wanted to counter the perception of the sheep farmers of South Tipperary as second-class citizens. The nominal head of the county sheep committee at the time was not a dedicated sheep farmer, and the committee of this IFA veteran was operating at a very low level of energy. To Ann, this explained why sheep issues were being neglected.

"For a couple of years, I fumed at every meeting at the insult to sheep farmers. Finally, a remark from the chair: 'Give the sheep farmers their 2 minutes,' made me really mad, and galvanised me into action. I decided to channel my anger constructively, so I ran against the incumbent chair of the county sheep committee. Following an energetic campaign, and against all the odds, I was duly elected chair."

Ann had zero experience in running for office, and even less experience in running a committee. But her trademark intense belief in the rightness of her cause, and the 100% support she enjoyed from husband Brian, drove her forward. A measure of the enthusiasm that Ann managed to generate can be gauged from the fact that 24 sheep farmers attended the very first meeting of her committee, 20 more than the average attendance hitherto.

"None of the people attending that meeting would have guessed how terrified I was at the prospect of chairing my first meeting. But I soon discovered that being chair did not mean having to be more efficient than everyone else. I realised that passion was more important than experience, dedication more important than knowledge."

Ann took to her new high profile role like a duck to water. She learned on the job, and proceeded to fulfil her election promise of bringing sheep farming into the IFA mainstream. As part of her job as chair of the South Tipperary Sheep Committee, Ann automatically became a member of the National Sheep Committee. When she arrived in Dublin to attend her first meeting, she was delighted to discover that there were more women on the committee than on any other IFA committee.

To the surprise of no one who knew her, by the time Ann left Dublin for home that same evening, she had become a member of the 8-person Management Committee. She threw herself into her new role with gusto, driving around the country in the red Mini that Brian's mother placed at her disposal. Brian did his bit by taking over some of her home roles, including taking care of Grace.

In the course of her work, Ann noticed that farmers in general have better attitudes to women, than the IFA does as an institution. "I am naturally proud that I became such a visible woman among a sea of men in the IFA. I gratefully acknowledge that sheep farming is gender-blind. I remember one occasion during an election, I accused one of the men of voting against me because I was a woman. 'I'll be honest with you, Ann,' he said, 'I never thought of you as a woman, good or bad!.'"

As a member of the Management Committee, Ann felt empowered to help make a difference about the things that mattered to her. "I loved the camaraderie that the IFA brings, and many of the men and women I met through the IFA became lifelong friends. I loved the rough and tumble of lobbying, cajoling and persuading, and the opportunity to rub shoulders with government ministers, TDs and Department of Agriculture officials."

Ann was happy hobnobbing with representatives of the print and electronic media, and she enjoyed her late night chats on her mobile phone (handsfree, of course) as she drove home from the interminable committee meetings.

Ann became actively involved in campaigning for Tom Parlon in his first bid for presidency of IFA. She campaigned across the country on Tom's behalf, and worked so hard that she actually lost weight. After the election, which Tom failed to win on that attempt, Ann had a well-deserved rest at home over Christmas. In early January, she travelled to Dublin for a committee meeting. She had been feeling a little queasy for a few days, but put it down to over-indulgence over the Christmas break.

When Ann's neighbour at the meeting lit up a cigarette (this was in the days before smoking became such a no-no), Ann suddenly felt positively unwell. "The thought occurred to me that I might be pregnant, but I quickly dismissed this as wishful thinking. For 14 years, I had so wanted another child, but I was mindful of my doctor's warnings. However, a visit to my doctor for a check up confirmed that I was indeed pregnant. I will never forget Brian's whoop of joy when I arrived back in the yard and told him the good news."

Ann wanted the whole world to know, and she positively blossomed during her unexpected pregnancy. When she attended her next committee meeting in Dublin, she gleefully shared her news with her neighbour, Mairead Lavery – who promptly divulged that she too was pregnant.

Ann's pregnancy was not easy, and she decided that once the baby arrived, she would give up her membership of all IFA committees. At the beginning of her ninth month, she chaired a meeting – and the next morning, three weeks early, her son David was delivered by Caesarean section.

Ann, Brian and Grace were overjoyed, but the next few months were very difficult. Little David was seriously ill. Ann was in the hospital day in, day out, for the entire period. And even when the baby was finally allowed home, he had to be tube fed for the next few months. However, after David had surgery at the age of 12 months, he started to mend, and the family started to function normally again.

After a break of 18 months, Ann was ready again to resume her IFA activity. "I was elected to represent all the farmers in South Tipperary on the National Council. At my very first meeting, under the presidency of new incumbent Tom Parlon, I met and befriended Jimmy Murray from County Roscommon. And it was together with Jimmy that I reached the highpoint of my career, when the two of us decided to take on the issue of sheep smuggling."

When conventional channels failed to get people's attention, Ann and Jimmy decided to go it alone in publicising the gross injustice of the illegal sheep smuggling that was rampant throughout the land. Ann had earned her activist spurs in the campaign against cheap New Zealand lamb flooding the Irish market, and she had also battled against the senseless bureaucracy surrounding the management of sheep dipping.

But sheep smuggling was different. This was a blatant attempt to undermine Irish sheep farmers, and this offended Ann's sense of right and wrong. She approached RTE reporter Charlie Bird and asked him to do an exposé on the subject. He agreed – on condition that they gave him proof. So Ann and Jimmy set out to obtain the evidence.

Armed with a secret camera, they photographed smuggled lambs being delivered at 4am to meat plants. After their first attempt at covert work, Ann discovered that her camera had malfunctioned, so they had to do it all again. This time they used hidden RTE cameras, and they caught the full extent of the smuggling operation. Ann and Jimmy handed their evidence to Tom Parlon, who gave it to RTE.

"The story broke one evening in November 1998, first on the 6pm TV news and then on the 9pm news. Overnight, I became a national celebrity. At first, I was universally vilified. The Minister of Agriculture practically called me a liar on national TV. But within a few days, the full extent of the scandal became clear, and my version was fully validated. I achieved almost heroine status among Ireland's sheep farmers. A few years later, when the foot and mouth crisis hit the Cooley Peninsula in 2001, my RTE footage was replayed on TV. Everything I had tried to warn people about was now coming to pass."

After the sheep smuggling exposé, Ann felt that she was ready to take on the chair of the National Sheep Committee. She wanted to be the first woman to achieve that position, and a couple of years later, she put her name forward for election. Ann lost the election by one vote. And even though she swiftly succeeded in being elected Vice Chair, the lack of confidence shown by her colleagues left her feeling let down, deflated and angry.

"At a personal level, it was a dreadful time. Just before the election, Brian's mother died, and immediately after, Henry Britton, who had first introduced me to IFA, was tragically killed in a car crash in Poland. My own father became ill, and died in July that year. I had been very close to my father, and the rapid deterioration in his condition in the months before he died affected me greatly."

The process of grieving, and the need to cope with all these emotional upheavals, led Ann to decide that she needed a break. She resigned as Vice Chair, on the grounds that if you can't do a job, you should allow someone else to do it. She also resigned from all her IFA commitments, and got to grips with her personal finances, upgraded her computer skills, and tackled some unresolved personal issues.

Finally, Ann felt able to return to the fray – but to a different fray. "My years in IFA gave me national recognition, and this proved very useful when I started seeking new pastures. I started to dabble in politics, and was appointed to the Advisory Committee of the Environment Protection Agency. Then I was invited to join the Council of the Grassland Association – where I found myself the only woman." Once more, Ann was given responsibility for an area close to her heart and her passions – sheep farming.

Looking back over her career, Ann feels blessed. "Brian has always been my stability and my inspiration, and I have thrived on his encouragement. He is a great man to have in my corner."

Ann is a very open, confident and liberated woman, and she constantly encourages more women to join IFA. She speaks from experience when she says that the insights of women are often sorely lacking when the big decisions are being made.

Ann herself has now moved on to new fields of endeavour. But everyone who knows her will agree that the farming world – or indeed, any other world that she gets involved with – has not heard the last of her. She has a remarkable circle of contacts and knows how to play the political game. If ever the IFA was to elect a woman President, I hope it is someone of Ann's calibre.

Theresa Wrafter

Theresa is like a kindred spirit to me. We seem to empathise on so much in relation to how we feel about equality and other issues. Theresa was involved in these matters long before I became IFA Equality Officer, and she wrote a letter to 'The Farmers Journal' years ago about the need to recognise women in farming.

Theresa has the gift of being able to give encouragement to others and to boost their confidence. She sends greeting cards to her friends, containing "believe in yourself" sentiments. Theresa is always willing to push out her own boundaries, and I identify her as a future leader.

The first thing that struck me when I met Theresa was the passionate way this feisty young lady expresses herself. When I pointed this out, she laughingly told me that she inherited this passion from her mother. Theresa still remembers overhearing a conversation between two of her aunts, as they were talking about their dear departed sister, Theresa's late mother.

"The aunts were reminiscing about my mother's habit of riding a man's bicycle around the locality. When neighbours were heard to comment that 'it was a shame for her to be doing so', my mother became even more determined that she would publicly ride that particular bike. I was raised and steered all my life in this same spirit of independence, and I am eternally grateful for the example I learned from my mother. If I am perpetuating my mother's passion for life, then I'm proud of it."

Born into a family of six children (four girls and two boys) in 1970, Theresa was reared on the 80-acre family farm at Ballinalack, County Westmeath. Although, or maybe because, the men were in the minority in the family, Theresa grew up with the comforting knowledge that the men-folk - her father and brothers – had a genuine regard for, and an appreciation of, women as equals. All the family were encouraged to appreciate what an invaluable role the tractor plays in farm life. Theresa was driving the tractor at an early age, and she still loves the challenge of mastering a new model of tractor.

Tractors also featured in an event that Theresa still regards as one of the high spots of her IFA career - the 2003 Tractorcade which started in Ballisodare and culminated in a massive rally in Dublin. Theresa believed passionately in the need to deliver a powerful message to the Government that it must never turn its back on farming and rural Ireland.

Theresa achieved the distinction of being the only woman to participate in the Tullamore Tractorcade protest. As she drove her tractor through the streets, waved on by enthusiastic crowds, she was proud of the strong show of unity displayed by the farming community.

Theresa loved growing up in rural Ireland and everything about the farm. From an early age, she developed a special adoration for the animals, and spent hours as a child, just loving and caring for the cattle.

There was never a dull moment for Theresa and her siblings. "My parents grew all our own vegetables, and I used to come home from school and spend many a cold autumn evening with freezing hands picking the potatoes my father had dug that day. We children helped with the hay, we saved the turf on the bog, and I can still remember the sunburn, as well as pains in my back and legs. My fun-filled childhood was a whirl of running around in the open fields, picnics on the grass, and working up a hearty appetite in the fresh air. The summers, in particular, were a time of enchanting freedom."

In the early days, Theresa's parents milked cows by hand, and as a young child, she helped with the milking until her tiny hands felt tired and lifeless. At some stage, her parents bought a bucket plant, which catapulted the farm into the modern era. Eventually, the dairy herd was wound down, and the family started a suckling enterprise.

Theresa attended secondary school in Mullingar, where she was the only farmer's child in a class of thirty pupils. She learned to hold her head high and spiritedly defended the farmers. This gained her respect from her urban peers.

From childhood into her teens, Theresa never seriously considered an alternative to a career in farming. "Yes, I successfully completed my Leaving Certificate, but my grades were not high enough to allow me to fulfil my ambition of becoming a vet. Any regret that I had at not working harder at my studies, was mitigated by the knowledge that I was going to inherit the farm. I suppose I'm a bit philosophical about my lack of motivation to shine scholastically at the time. But I am happy with my choices, and feel proud and privileged to be a landowner and a farmer."

Between leaving school and committing herself to farming full-time, Theresa took time out to travel. She worked as an au pair in England and Canada, and then went to New York City to work for a year. The Big Apple had a captivating appeal, and she was almost tempted to stay. To this day, she retains a deep fondness for the United States, and her dream is to one day buy a holiday home in Vermont, and retire there for six months of the year.

While out socialising with her friends, Theresa used to amuse herself by playing a little game with men she met, asking them to guess her occupation. Nine times out of ten, they would say she was a model. Emboldened by this, she entered the Rose of Tralee contest in 1991, and was beaten in the finals of the Miss Westmeath contest by her best friend. Theresa sometimes muses on how unusual it would have been for a farm girl to become the Rose of Tralee.

Soon after returning to Ireland, she found love, in the shape of Kevin. "We actually met at a dance in Tullamore, but Kevin could not muster up the courage to ask me to dance. However, he boldly followed me home with his friend, and eventually we ended up talking into the small hours of the morning. I had never seriously considered settling for anyone who wasn't a farmer, so it was no coincidence that dairy farmer Kevin fitted the bill. I fell for his spontaneous nature, and I'm happy to report that I have enjoyed an eventful, joy-filled marriage for over nine years with our children Fabian, Denise, Ciara and baby Gearoid."

Theresa farms alongside Kevin, and between them they own 240 acres of land. They work as total partners, mainly dairy farming with a 365-day supply. They hold a quota of 100,000 gallons of primarily liquid milk as well as a suckler herd. Luckily, the things she dislikes, Kevin likes – and vice versa. This makes for a good match, and a smoothly run business.

Earlier experiments with hired labour were abandoned, when Theresa and Kevin found that employees showed too little respect for machinery and livestock. As a result, they made a conscious decision to work harder themselves.

Theresa has acquired an intimate and absolute knowledge of dairying. "I know that dairy cows can be very demanding creatures, and are under enormous pressure to produce milk. They are vulnerable, need careful daily attention, and are prone to falling prey to weakness. Dairy farmers handle their cows with more attention than other cattle farmers, and the work can be quite stressful at times. Bringing in cows to milk in the pouring rain can be a torturous chore, that takes its toll on both man (or woman) and beast. I know our dairy cows like the back of my hand, and I can detect illnesses in their early stages. I constantly monitor them for milk fever, tetany, mastitis, listeriosis and a multitude of other health problems."

For Theresa, each cow is an individual. She claims that if animal rights activists spent a year caring for every aspect of a herd's upkeep, they would gain a better understanding of farmers.

Although Theresa admits that milking can be monotonous, she never finds it boring. As the cycle of the seasons come and go, so too do the various activities within the herd. Her favourite time of the year is spring/summer. She loves seeing the cows go out to grass in early spring, after a long winter indoors.

Heifer calves are the start of the dairy herd chain. "I find that training 30 to 40 baby calves to drink milk from a bucket each year can be an emotional job. I get deeply involved with these calves, which are valuable both in monetary and genetic terms. I believe that it is wiser for a calf to drink milk as opposed to sucking a man-made teat. The sucking tendencies nurtured by these teats remain with a heifer, which can develop into a dreadful habit of sucking her peers, one of my pet hates."

A cow coming into the parlour after calving is a moody animal. She is stressed, her udders are sore, and she tends to literally kick the hands off you. "As a woman, I find it easy to understand cows, having experienced some very painful birthing realities!"

Theresa has always been fascinated by how children's minds can be moulded by contact with nature and animals. Her interest in this led her to gain a Montessori diploma, which proved a great asset in the rearing of her own children. She believes that the basic instincts of humans are natural and tactile. Unfortunately, according to Theresa, urbanisation is turning humans into stifled, mechanised beings.

Theresa claims that farmers are almost the last remaining members of society who really understand nature. "I was shocked to discover urban kids who believe that milk comes from a carton or that chips come from a chip machine. I was so frustrated that the milk consumed by the public seems so far removed from the cow that produced it, I decided to become involved in Agri-Aware, an organisation that strives to achieve a balance between the urban and rural divide."

As part of an Agri-Aware pilot scheme entitled "Bringing Agriculture into the Classroom," Theresa gave one-hour presentations to classes of schoolchildren, educating them on how their food is produced, and teaching them that farmers play an important role in society as the custodians of the landscape and caretakers of the environment.

While visiting schools to deliver this programme, Theresa saw that even in rural schools, there were usually not more than four farming children in a class of thirty. "This is a sad reflection on how urban living has smothered the rural farming base." However, she is greatly encouraged by the enthusiastic and positive response of the children during these class visits.

Maternity leave is not a concept readily understood on a farm, so Theresa continued farming throughout her four pregnancies, which meant working right up to full-term. Milking cows, driving the tractor and feeding the calves helped keep her fit and energetic. Kevin long since gave up trying to tell her to take it easy. He knows that his wife is not someone who likes being told what to do.

But even Theresa's remarkable fitness could not help her when she developed deep vein thrombosis after the birth of her second child. An overpowering yearning to be with the baby meant that she ignored the painful swelling in her left leg. She was eventually persuaded to visit a doctor, where she discovered that a clot was travelling to her lungs. The GP harshly informed her that she was just hours from death, and rushed her to hospital.

"I spent the next six months on blood thinning medications, and I also managed to suffer appendicitis. But I was soon back working on the farm, maybe even sooner than I should have done."

The inferior status of many women in agriculture occupies Theresa's mind. I asked her to give me an example.

"One winter morning, I was feeding young cattle in a shed with round bales of silage, while Kevin was spreading slurry from an adjacent shed. As I drove the tractor around the laneway, laden down with a round bale on the front-end loader, a car drove up. The driver first approached the front door, but when she spied my tractor coming round the lane, she made a beeline for me. The visitor, a representative of an agricultural association, asked me to direct her to Kevin. When I had the temerity to seek the nature of her business, she stuttered something about needing to talk to Kevin. I waited for her to enquire whether I might be a decision-maker on the farm, but she continued to stare at me. I regarded this as the ultimate insult. It's bad enough when a man fails to recognise me for what I am, but a woman!"

This incident made Theresa more determined than ever to work towards a change in attitudes. She has very firm views on the question of inheritance, insisting that her two daughters have as much chance of inheriting one of the family's two farms as their brothers. It will depend on whether they see farming as a career choice. "Equal opportunity" is Theresa's mantra. If a young girl is interested in farming, she should inherit. Land ownership is too important to leave the women out of the picture.

This young woman has an ambition that her present journalism studies will one day reap as much fulfilment as farming. Theresa finds expression through the written word an effortless and rewarding exercise.

In 2004, Theresa reached the final shortlist of the 'Woman's Way' Mother of the Year Awards. The judges included RTE's John Creedon and TV3's Lorraine Keane. After being featured on RTE's "Off The Rails" because of her career in dairy farming, Theresa admits that the whole experience was a "fleeting moment in time, but equally a highly memorable one."

Theresa firmly believes in the importance of positive media attention on agriculture, because it facilitates an understanding between the urban/rural divide.

"I wish to graciously attribute my new role as County Secretary for Offaly IFA after just three short years with the association, to the confidence-boosting qualities of

Mary Carroll. It is due to Mary's enthusiastic encouragement that I am determined to continue as an ambassador for women's involvement within the IFA."

While Theresa's children and family are who she adores most in life, farming remains her passion and the simple sustenance of her very being.

Elizabeth Hogan

Elizabeth was one of the first women I met when I started as the IFA Equality Officer. I still remember accepting an invitation to meet IFA women in the Abbey Court Hotel in Nenagh. They were so kind and I felt really encouraged. I knew I was forging a new strong link for the future.

It was from Elizabeth that I learned a great quote "Instead of seeing the rug being pulled from under us, we must learn to dance on a shifting carpet." That's Elizabeth all over, always ready and positive to meet the challenge. She proved it when I couldn't find any other woman driver to take on the task of driving the tractor during the Tractorcade.

Elizabeth describes what happened when I asked her to contribute her story to this book.

"One sultry August day the combine harvester had broken down yet again in the cornfield. The men folk were none too happy, and stood around muttering. Just then the postman arrived. His bright, cheery smile and quick wit succeeded in lightening the heavy August air. He delivered a letter to me. The letter was from IFA Equality Officer, Mary Carroll. I knew her as a dynamic young woman who had enhanced the lives of so many farmwomen since she first took up her post in 2001.

"Getting a letter from Mary was not an unusual or unexpected event. Maybe this was the latest Agri report? Perhaps it was more information about training courses and conferences. Or could it be an update on the forthcoming trip to Australia? But no. I was dismayed to discover that, instead of containing Agri info, the letter brought a request from Mary asking me to provide information about myself for inclusion in a publication about farmwomen. My first reaction was 'God above, what's interesting about my life?'"

But with some gentle persuasion, Elizabeth finally relented and consented to be included in the book. She is the youngest of three children who were raised on a farm near Lough Derg. Her memories of her early childhood are happy ones. This was an idyllic period, with plenty of time for fun and freedom. As she grew up, she realised that her parents really valued the most important things in life, things that no money could buy.

"Sunday was always a special day. Each Sunday evening the family would depart for a surprise drive in our old Ford car, registration number FI 6394. Often accompanied by Aunt Lizzy, we would be ferried around Ireland by Mam and Dad who turned the trip into a combined history and geography lesson."

Once Elizabeth finished primary school, the natural progression in her home was to head off to boarding school, just as her sister Margaret and brother John had done. As September approached, preparations were made, bags were packed, and Elizabeth was enrolled into first year of the Brigidine Convent in Mountrath.

"Boarding school life in the fifties was much stricter than today. There were no weekends home, and everything happened within the high stone walls of the convent. We all arose at 6.30am for oratory prayer and daily mass. When I look back at that time, I remember the nuns as fantastic educators. Not only was there the daily school curriculum but also music, drama, gym, elocution, sport and constant preparatory sessions for life. In retrospect, I realise that as teenagers, my siblings and I probably failed to appreciate the foresight of our parents' decision to ensure that we received the best education. Nor did we appreciate at the time, the huge sacrifices made by our parents to provide for this education."

It was during her boarding school days that Elizabeth first met the young man who was to become her husband. Their friendship, and eventual courtship, started when Billy, with some friends, attended the annual November opera production.

Poignantly, it was also on a freezing cold November day in 1962 that life was to take a dramatic turn.

"Mother Superior summoned me to her office and gave me the devastating news that my beloved father had died suddenly at the age of 55. This was not the only trauma that I experienced while at school. Exactly three years later, on the anniversary of my father's death, my mother died. Ten months later, Billy's sister Margaret, who was a member of cabin crew with British European Airways, was killed in a car crash in Britain. Still only a teenager, I had to come to terms with the harsh realities of life and death."

Elizabeth recalls her father, Joshua, attending meetings of the IFA (then the NFA) in Nenagh. Names such as Rickard Deasy, Paddy Kennedy and T.J. Maher were familiar to her from the chat at home. In September 1966, shortly after Elizabeth and Billy were married, the now famous farmers' march took place, when farmers walked from Bantry to Dublin. The march was led by NFA President, Rickard Deasy, a native of North Tipperary.

Elizabeth remembers him as an exceptionally strong leader, both physically and mentally, and she can still see him clad in leather boots, black tam and carrying a blackthorn stick. He led frustrated and weary farmers to the doors of Dail

Eireann. As they marched from County Cork to Dublin, they were joined by supporters from every town and village. Nenagh was a particular stronghold of support. When Elizabeth and Billy returned from their honeymoon, they helped out with medical supplies and food. Their neighbour, Paddy Kennedy, proved to be a superb organiser and assigned duties to everyone. Elizabeth's most vivid recollection of this time is the blistered feet of the marching men.

"During the 1960's, the involvement of women in the NFA tended to be more low key than today. But Sheila Deasy, a most gracious lady and friend, sticks out in my memory. She was a model of emotional and physical support for her husband. Day after day, she sat beside him on the steps of Leinster House, in the freezing cold, quietly doing her knitting. It was clear how important her role was as his closest friend, advisor and confidant, whilst he tried to restrain the angry farmers. I would like to believe that if a similar march were organised today, farmwomen in general, and IFA women in particular, would be out there marching side by side with the men."

Elizabeth and Billy built their home on Billy's family farm. His father did the daily herding and had a liking for pigs. His mother mostly did her housework and tended to poultry. She also supplied eggs and milk to her near neighbours on a daily basis.

"Our eldest son, William, was born in 1967, followed by Brian, Joseph, Helen and David. Within a week of Helen's birth, both my uncles died suddenly. Just over seven years later our youngest child David was born."

As Elizabeth and Billy were raising their young family, they also built up a dairy herd, constructed farm buildings, milking parlour, slatted sheds and silage layout. This was a very busy, but also very fulfilling, time. Elizabeth had to cope with helping to run the farm at full production, while also providing for their own home, and the home of Billy's parents.

For Elizabeth, the fact that she and Billy always worked so well together as a team made a big difference and was an added bonus. When possible, they managed without outside farm labour. If an emergency occurred, either inside the home or out on the farm, they dealt with it together.

"Throughout my life, births and deaths, celebrations and tragedies, seem to have gone hand in hand. Billy's father died at the age of 83 and his mother died just before David's birth. This was also the time when I started experiencing a reoccurrence of an old hip problem. I had been born with a congenital defect, and as a child I had been under the care of an orthopaedic surgeon. At that time medicine was much less advanced than today and surgery was considered inadvisable. My hip had become arthritic and I had started walking with a pronounced limp.

One could only liken the searing pain of arthritis to a hot poker burning into the bone and it was often worse in the heat of summer than in the cold of winter."

Elizabeth decided to put off any surgery while the three oldest boys were still studying at their respective universities. Once she had seen them through their college courses, she finally underwent her first hip replacement at the age of 43.

"I so appreciate just how great Billy and the family were during this period. Each member of the family offered his or her own particular strength. The surgery gave me a whole new lease of life and five years after my first operation I had a second hip replacement."

Elizabeth certainly has a rather philosophical attitude to pain. "It can sometimes be so intense that one can actually go beyond it. If you have been there you will know exactly what I mean."

She values the academic prowess of her children, all of whom studied at university. William qualified in Medicine at Trinity College Dublin and is currently a consultant at the Mayo Clinic, Minnesota, USA. He and his Dublin born wife, Gwen, have three children, twins Oscar and Emily and baby Naomi. Brian qualified as a veterinary surgeon and he and his wife, Vicky, live in England and have a son Fintan and a daughter Rose. Joseph studied Food Science at University College Cork and works in the high tech industry. Helen studied Journalism at Dublin City University and at Humboldt University in Berlin. After working as a journalist, she later qualified as a teacher at Mary Immaculate College, Limerick. David, the youngest son, recently graduated from the University of Limerick.

Now that all the children are beyond their college days, Elizabeth is happy that books will always be part of her home. "I recall with great happiness the sheer fun and enjoyment I experienced as I watched my family grow. I have treasured memories of graduations, weddings and christenings. I cherish special occasions such as the birth of my grandchildren, the beautiful music at William and Gwen's wedding, and the impromptu Ireland v England soccer match held on the lawn of a beautiful country house in Bristol at Brian and Vicky's wedding."

Elizabeth never forgot how involved her father had been in supporting direct action by farmers. During the 2000 beef protest she, and other IFA women, organised food and kept night vigil with the men. The January weather was particularly glorious. She remembers sitting at dawn on bales of straw at an open campfire enjoying the heavenly chorus of birdsong from nearby trees piercing the crisp morning air.

During the protest, Elizabeth saw much evidence of unity and friendship among farmers. She believes it is wrong to ever doubt or underestimate the basic goodness of people. Public opinion was at an all time high with hundreds of supporters flocking to the factory gates every night to hear the latest report from the negotiating panel and to partake of refreshment. There was tremendous generosity and goodwill among the local community. Business people and professionals, as well as people with no direct connection with farming, supplied food, drink and support.

Later in 2003 Elizabeth also became involved in the IFA Tractorcade. The tired tractor drivers stopped to rest at Dunkerrin Hall, near Nenagh, where Elizabeth and her IFA colleagues supplied refreshments.

On this occasion Elizabeth was appointed to drive a tractor in the convoy. However, her own home tractor, a Ford 7610, had clocked up many miles over the years and she didn't feel confident that it could reliably make the journey. Into the breach stepped Bobby O'Brien, of O'Brien Brothers, Carrig, agricultural contactors and machinery geniuses. Bobby offered Elizabeth his tractor, a New Holland 8360, 145 hp. It was like a computerised space machine, but after a quick lesson from Bobby, Elizabeth mastered the tractor. To quote Jack Charlton about his soccer lads, "We got a result".

As Elizabeth surveys the changing nature of life on the farm over the years, she remains adamant that farmers can and must change with the times. She sees this as a time to learn to dance on the shifting carpet and she is a great believer in the maxim "Go out on a limb, that's where the fruit is". For Elizabeth, this is the time to look to the future and to spot the new and wonderful opportunities on offer.

Marie Doherty

Marie Doherty is a real woman in agriculture, and her attachment to the land is palpable. I will never forget how much she missed the feel of the grass on her feet when we were together in Australia recently. As soon as she saw a field, she ran straight to it.

Marie is not only the youngest farm woman to be featured in my book. She is also the youngest ever County Secretary in the history of the IFA. She was a mere 19 years old when she was elected as Donegal IFA County Secretary in 2002, and she also serves as secretary for her local branch and for the Inishowen region.

Marie was raised on a small Donegal suckler farm in Malin Head. As the most northerly point in Ireland, Malin Head is a name known to anyone who listens to the radio, being mentioned several times a day on the weather forecasts.

"I am an only child, and one of my earliest memories of growing up on the farm is of following my father around the farm. I used to pester him with an endless stream of questions about every aspect of farming. Why did this happen? What was the purpose of that? Why did this need to be done? How did that work? I had an unquenchable need to know everything, and I am probably still driven by a need to know what makes the world – and especially the farming world – go round."

Marie's neighbours will all tell you that her father was always in such a hurry that he never took time off. However, when it came to Marie, he seems to have been blessed with infinite patience. He appreciated his daughter's almost insatiable thirst for knowledge, and was eager to share whatever information he could with her.

The man regarded by everyone as a workaholic always made time for Marie. There was also a healthy dose of self-interest involved. By taking out a little time now and again, he hoped to reap the benefits in the future.

"Given the sometimes fractious relations that can exist in Irish farm families as a result of inheritance issues, I sometimes wonder if my father's attitude would have been different if I had a couple of brothers. I have even discussed this with my father, who always assures me that even if there had been a larger family, I would have had an equal opportunity to participate in the farm work. I have no difficulty accepting this. From the time I can remember, equality reigned at home. Anyone who ever visited the family farm was expected to muck in and do something useful. If any of the visitors showed any reluctance to offer their help, there was always someone on the farm to tell them in no uncertain terms what was expected of them."

Marie is sure that even the presence of brothers would not have diminished her equal status. Nevertheless, if she is entirely honest with herself, she acknowledges that there would have been at least one beneficial side effect of having boys around the house.

"I might have developed a slightly more conventional attitude to agriculture. By conventional, I mean more feminine. Maybe I would have allowed myself to devote a bit more time to traditional skills such as baking scones. Not that I'm complaining. On the contrary, I greatly value the fact that as the only child, I simply had to master the more practical end of things."

As Marie grew up, she developed a range of farm skills that were to prove very useful later. And it's not every farm woman who knows how a tipping pipe works, how to hook on a linkbox, or how to tag a calf (not to mention knowing why the calf had to be tagged in the first place!).

"My mother, Josephine, was the second oldest of seven children born in the nearby town of Carndonagh, where she worked in a shirt factory. Although she was not from a farming background, she fell in love with and married a farmer, and soon grew to love farming. My father started working with Donegal County Council as a roads labourer. Like thousands of other farm women, my mother took on much of the unseen everyday work on the farm. I am still in awe of how she adapted to the unfamiliar farm surroundings. I regard my mother – together with all the other farm mothers – as one of the true heroines of Irish farming."

The family has had its share of heartbreak. One of Marie's uncles had died of cancer in 1999, and another uncle died in his late twenties while working with the Department of Agriculture. Her grandfather died in 2002. But nothing prepared Marie and her family for the tragedy that was just around the corner.

"In March 2002, the day before my twenty-first birthday, my mother Josephine was diagnosed with cancer. The family had six short weeks to get used to the news before she succumbed to her terminal illness at the young age of 52. It is at times like these that the real value of rural community can prove itself. The house never emptied of neighbours and friends over those first days. I know that my father and I could never have got through it without them. There was always someone there to do anything that needed to be done."

The loss of her mother was a great shock to Marie, and pushed her to examine her life priorities. It was only after her mother's untimely death that Marie - and her father -realised just how spoiled they had been by Josephine. They also realised that they may have taken her too much for granted.

Marie is sure that the IFA camaraderie helped her get through that difficult time. A few days after the funeral, Marie was back at the County Executive, grateful for some sense of normality in a world that had fallen apart. The fact that she had to continue attending meetings helped show her that life does go on, even when it seems that the light has gone out.

"This social aspect of IFA may not be listed as one of the official benefits of membership, nor has it been given much prominence as the IFA celebrates its fifty year anniversary. But I firmly believe that most IFA members would agree with me that this social aspect is one of the organisation's major strengths. For me, one of the biggest benefits of being involved in the IFA is the number of people I have met. Thanks to the IFA, I now have friends from all over the country."

The question of whether to join the IFA was never if, but when, for Marie. She had always had a keen interest in farming, and had always assumed that it would be a good idea at some point to become a member. Marie knew that farming would feature prominently in her future, so it made sense to become a member of the country's largest representative organisation.

Another reason why Marie always assumed that she would join the IFA was that membership would help fill her need to store knowledge, a need that had first manifested itself as she followed her father around the fields as a child. She was eager to learn more about the changes that would affect the farming community in general, and her own future in particular.

"The actual impetus that prompted me to join the IFA was the impending IFA presidential elections in December 2001. I felt that I wanted to know more about the candidates seeking to represent the IFA at national level, so I joined in October before the elections. I had not realised how many new people I would meet, and I had not bargained for the fact that membership of the IFA would lead to such a rapid expansion of my circle of friends and acquaintances."

Marie took on the secretaryship of the local branch, and then became Donegal County Secretary in March 2002. She says proudly that the Donegal lads could not have made her feel more welcome. In the words of one of the executive officers on the County Board, the lads "are 100% behind you."

It became apparent that the huge geographical area of Donegal was impeding communications within the county. In order to streamline the flow of information, it was decided to add an extra layer between the local branches of the IFA and the County Executive. Nine branches on the Inishowen peninsula got together to form the Inishowen region. Naturally, the person chosen to be secretary of this new body was Marie. This region now holds regular meetings, and the system has worked well.

Marie takes a very matter-of-fact approach to the fact that she is so capable in the various posts she holds despite her young age. She has proved by example that, as long as you can do the job well, gender or age should make no difference. Marie is particularly proud that a Donegal woman, Mary Coughlan, was given the top post in Irish agriculture. Shortly after her appointment as Minister for Agriculture, Minister Coughlan attended the annual Inishowen IFA dinner dance.

"I believe that for anyone involved in farming, it should be natural to gravitate towards the IFA. It makes sense to want to support the organisation that has represented Irish farmers for fifty years, through good times and bad. I would encourage any young farmer to contribute their new ideas and skills to the IFA. I would also encourage older members to contribute their wealth of knowledge and experience."

It was through her active involvement in the IFA that Marie got to hear about the 'Farm Hands Across The World' trip to Australia. She was the youngest of the group of thirty-two women who packed their bags, hopped on a plane, and for three weeks travelled along the east coast of Australia.

"One of the weeks included a wonderful stay on a farm. The family that hosted me opened their home and their hearts, to take me in and make me feel like one of them. I think I learned as much about Irish women on the trip as I did about Australian agriculture. I loved the opportunity to meet the other members of the group, and they all served as excellent goodwill ambassadors for Ireland. It was fun being the baby of the group, and I'm really grateful that the others all took such good care of me."

Marie continues to punch way beyond her weight, and is already making an impact in the IFA far beyond her years.

Nora Duffy

When Donegal sheep farmer Nora Duffy opens her mouth to speak, you might expect to hear some variation of a regional Irish accent. Instead, the words that pop out have an unmistakable and delightful Scottish lilt. That's because even though Nora's ancestry is impeccably Irish, she was born in Scotland, and spent her formative years there.

Nora has always struck me as just being someone different. Of course, the accent has something to do with it, but also that fact that she came into farming from a totally different background. I admire her ability and willingness to come forward and get-involved in the IFA. Nora is a strong, outspoken woman. I love her confidence, and the bohemian attitude she brings to farming.

Both Nora's mother and maternal grandmother were Scottish of Irish descent, while her maternal grandfather was from Connemara, County Galway. Nora's father and his parents were from Donegal.

"As a child, I used to spend every single summer holiday in Ireland. For half the summer I would be with the Galway branch of the family, and for the other half I would be with the Donegal branch. I have it on very good authority that, it was while I was spending time during the Donegal half of one summer, I took my very first unaided steps in life. The Donegal air has always seemed to inspire me."

Born in 1954, the eldest of four children, Nora had married very early. She embarked on nursing training, and at the same time she worked in several hospitals, both in Scotland and in England. It was during this period that she gave birth to her son in 1980. Unfortunately, she was experiencing problems with her marriage, and when her son was two years old, Nora and her husband decided to divorce.

In the meantime, Nora's parents had retired, and moved from Scotland back to Donegal. Nora decided that this was where she wanted to raise her young son, so she left Scotland and joined her parents and her two sisters in Donegal. Here she set out to start on a new life in a new, but not totally unfamiliar, environment.

A couple of years later, Nora met Brian, and they eventually got married in 1987. Nora and her son moved into the farm where Brian and his mother lived. For Nora, this was her first real experience of the farming way of life.

Prior to marrying Nora, Brian had established a regular pattern of spending the winter months with his two sisters and two brothers in America. This continued even after he married Nora, and each autumn, Brian continued to make his annual visit to the States. It was during one of his absences that Nora made an unpleasant discovery one winter's day. She was shocked to find that four of Brian's sheep had developed Pink Eye, or 'Dullnamullag' as it is known locally.

"Given my relatively little experience with sheep farming, it was clear I needed help. I asked a local lad to put the sheep into the byre where I intended to treat them. He shared his expertise with me, and told me that the best way to heal Pink Eye was to put powder into the eyes of the affected sheep. This did not seem too difficult a task. I suppose I thought that I could happily draw on my nursing experience. So I was quietly confident that I could manage to treat the four affected sheep. The whole task did not fill me with too much trepidation."

However, problems arose as soon as she looked at the sheep. She realised that while seasoned sheep farmers may be able to tell one sheep from another, she certainly could not.

"What was I going to do? I racked my brains, and then I came up with what I felt was a novel and effective solution. Using a permanent marker, I wrote XOXO on the horns of each ewe I had treated. I was probably feeling pretty pleased with myself, and I got to thinking that with my system, I would have no problem the following day. I was right. Next day, I was able to distinguish which sheep had received the powder and which sheep had not."

What Nora had not bargained for was the reaction of her neighbours. They could not believe their eyes when they saw what she had done to the ewes' horns, and the incident continued to be a source of great merriment throughout the district for a while. But Nora was undaunted. She was proud of her exploits. She was delighted that even though Brian was away, she had actually enjoyed handling the problem on her own.

What the horns incident also did was to kindle Nora's interest in sheep. "I felt it was time to familiarise myself better with sheep farming, and I was determined to try and become more experienced by the time Brian returned from his trip. Every evening I set out, accompanied by the dog. But this brought its own set of challenges. Brian was a natural Irish speaker, and he had only ever spoken Irish to the dog. It did not take me long to grasp that my canine companion did not understand a word of what I was saying. The poor dog could only respond to instructions to gather and check the sheep when these instructions were delivered in the Irish language. Even if the dog had learned rudimentary English, he had certainly not mastered English with a strong Scottish twang."

So Nora embarked on a campaign to teach the dog English. Slowly but surely, the dog started responding to Nora's hand signals. By the time spring came round and Brian returned from his travels, he discovered to his amazement, amusement and bemusement that Nora had successfully trained the dog to understand English. They were now the proud owners of the only bilingual dog in Ireland!

Nora grew very attached to her smart sheepdog, and when he died, Nora bought two pedigree Border Collie pups. She undertook all the training on her own, and to this day they make a rare team.

"Encouraged by my experience with the sheep and the dog, I gradually began taking on more and more of the day-to-day running of the home farm. By 1993, I felt proficient enough to agree to Brian's suggestion that the flock number should be registered in my name. There was not much difficulty in organising this, since the land was already in our joint names."

For many years, there was an assumption in the family that Brian's American brother and his wife would one day return to Ireland and run their own farm, which Nora and Brian had been living in, and were looking after for them. But in 1996, they announced that they did not intend to leave America, and they gave Nora and Brian their farm. When Brian's mother died, Brian's other brother took over the home farm, but it was Nora and Brian who undertook to care for this farm also.

"It was in a dreadfully dilapidated state of neglect, all covered with rushes, thistles and bracken. Together with my husband and son, I decided to take on the task of clearing all four fields manually. It was hard work, but we enjoyed it. I still remember the huge sense of satisfaction I had when I could declare the fields clean."

Soon afterwards, Nora decided that it was time for the family to have a new home. She personally involved herself in the minutiae of drawing up the plans and in supervising its construction. She chose to design a modest house that looks like an old cottage. Visitors who arrive find it hard to believe that the building is only a few years old, rather than something built a hundred years ago.

For ten months, Nora, her son, Brian and their friend John, all participated in building the house. Nora herself helped with the plastering. Their hard work was rewarded, and they were relieved and delighted to move into their new home on St. Patrick's Day 2000.

A year earlier, Nora had started experimenting with pedigree rams that she let run with her Cheviot/Mountain ewes. "I had come to the conclusion that in the medium and long term, demand for my sheep would decline, because they were not reaching the right weight. I decided that Texel ewes were the answer, and as soon as we completed our new home, I bought my first pedigree Texel ewes."

Nora was starting to be an acknowledged sheep expert in her own right. She joined the Texel society, and she is now starting to get good results. Nora believes that the day is not far when she will be able to offer pedigrees for sale. All this is in addition to, and not instead of, Nora's commercial flock, which she has no intention of giving up.

As Nora became more familiar with agriculture, she started attending county IFA meetings. She enjoyed the experience of mixing with other farmers, and in 2002, she was asked to take the chair of her local IFA branch. She had fun during her term of office, and her only criticism is about attendance levels at meetings. "It was disappointing to discover that too many farmers are still too apathetic about their representative organisation."

In addition to developing her expertise in sheep farming, Nora has also found another outlet for her creativity.

"I started developing a passion for painting, and I indulge in this leisure pursuit with three friends, and we all paint together. One of the foursome is my best friend, who happens to be a small animals vet who treats my dogs and cats. This friend's husband is also a vet, and he treats my large animals. Every Tuesday evening, we meet at the home of Patricia Mann, who formerly worked as an art director in America for many years. Patricia is now a fine arts painter residing in

Donegal. We have a meal together, enjoy the craic, and then embark on some serious painting until the wee hours. I just love painting in the company of my friends. I find it very relaxing and therapeutic."

In October 2004, Nora and her friends held their first exhibition in Donegal Town, where a Scotsman turned the upper floor of his shop into an art gallery. Coaches of visitors stop to view the paintings, and Nora is delighted that her art has sold nicely. The friends held an exhibition in the Gweedore Credit Union, and exhibited in the Gweedore Credit Union. Nora had one of her paintings accepted by the prestigious Glebe Gallery, of Derek Hill fame, and there are plans for further exhibitions.

Nora's passion for painting has in no way diluted her passion for farming. "I look forward to being able to continue to farm for many years to come. If the future is anything like as fulfilling as the past, I will be a very contented woman."

Kate McMahon

Kate was one of the first women to contact me to offer support and help when I took up my post as Equality Officer of the IFA. We chatted a lot on the phone, and she was always pushing me to achieve even more. As a very talented writer herself, she encouraged me to make greater use of the media to spread and promote our aims. As someone who often wonders whether or not I want to farm, and where I should start if I do, Kate has been a total inspiration.

Here is Kate's story in her own words.

"I was born in Boston, Massachusetts in 1950, and grew up in a small town twenty miles west of the city. I am one of four daughters, and the younger sister of a set of non-identical twins. Both my parents are college graduates, and made it quite clear that they intended that their four girls do the same.

However, I found school difficult from the very start. I remember in first class having to copy answers from the boy who sat next to me, and was so desperate the one day he was absent from school, that I ate the powdered soap in the girls' toilet so that I could be sent home sick as well. I loved hearing stories read aloud, but disliked anything to do with reading, spelling, and writing. When I was in fourth class, I was diagnosed as dyslexic and started receiving remedial help.

My first memory of farming occurred in kindergarten when I was about six years old. Our teacher gave each child in the room a sealed jar with a small amount of cream in it. She told us to shake the jar as hard and as long as we could. When we opened our jars after shaking them, we discovered that our cream had turned into delicious butter that she spread on cream crackers for us to eat. I will never forget the feeling of delight of producing something with my own hands.

My second memory of farming was when I was a few years older and my twin sister and I would study the 'pet section' of the Sears and Robuck mail-order catalogue sent to our house each year. A small grey donkey was listed as one of the kinds of animals for sale and we begged our parents relentlessly to buy us one. I remember spending hours pulling the long grass that grew along the edge of our garden and drying it in the sun to make hay just in case my parents gave in and bought us the $99.99 pet. But they never did.

The next decade of my life was coloured with a never-ending stream of household and not-so-household pets. The list included buckets of tadpoles, tanks of goldfish, painted turtles, green lizards, school hamsters, tiger-striped cats, and Labrador retrievers. Each animal would be dutifully christened upon its arrival and then solemnly buried in the back garden when its time came to an end.

By the time I was in secondary school I knew I wanted to help dyslexics avoid the frustration and humiliation I suffered as a child. I decided, regardless of how long it took, to become a language disability therapist. I graduated from the University of Vermont in 1974 with a degree in Elementary and Special Education. I went on to attend Lesley College in Boston where I received my Master's degree in Reading in 1976, and I picked up another degree in Developmental Dyslexia from Massachusetts General Hospital in Boston.

In 1982, I moved to County Galway to set up a teaching practice for dyslexics. It was only then that I finally managed to acquire my first large animal; a six year old, strawberry roan mare I named "Haley" whose offspring I still continue to breed from.

In 1986, I married a local farmer, Gerry McMahon, from Craughwell, and the agricultural side of my life truly blossomed. Gerry ran a mixed herd of horses, suckler cows and sheep to which I added pigs, turkeys, ducks, geese, and hens. I spent the next seven years of my life happily raising our two children, Amy and William, and learning as much as I could about animals and land.

In addition to helping Gerry look after the farm, I grew fruit and vegetables for our own consumption and froze, preserved, and pickled everything in sight. My greatest pleasure was producing a dinner that was completely home-reared and grown; one of my favourite menus was cream of parsnip soup, roast pork with applesauce, new potatoes, asparagus, and rhubarb crumble.

In 1993, Gerry suggested that I buy a small farm that had come up for sale four miles away. I discussed it with my family at home in the States and gathered up enough money to make the purchase. Gerry helped me select the stock, twenty five Limousin heifers and an Aberdeen Angus bull.

I loved the opportunity and responsibility of having my own animals and land to look after. I applied for a herd number, joined REPS (Rural Environmental Protection Scheme), consulted with Teagasc (Irish Agriculture & Food Development Agency), and became a member of the IFA. A friend who was visiting me at the time from the States suggested I write a children's book about my life on the farm. My immediate response was, 'I can't write, I'm dyslexic!'"

But after my friend left, I thought about it and how much I loved having stories read to me as a child and I decided that I had nothing to lose trying to write some-

thing except a bit of my own free-time. I wrote a book using a diary format about the summer of 1996, when our daughter Amy was loaned a pony named Timber Twig to compete in the local shows. 'Timber Twig', with a lot of help from the editor, was published by The Children's Press in Dublin in 1997.

Then, in the same year, thanks to the economic boom of the Celtic Tiger, and help from the bank, I rolled my farm over into a much larger holding in Currantarmuid, Athenry. I expanded the herd to forty sucklers and bought seventy Suffolk breeding ewes as well as keeping my broodmare.

I wrote a second book about our son William, 'Horse of Another Colour', in 1998. In both the years 1999 and 2000, I was selected runner-up in the Bank of Ireland Farm Woman of the Year competition. And then in the year 2000, I wrote a third book, 'Growing Pains', and was selected in 2001 as Regional Winner in the final year that the Farm Women competition was held both books were published by The Childrens Press in Dublin.

My farming philosophy has always been simple; "what I do, I do well". I also believe in being very open to new ideas and suggestions, and am always asking questions and observing how other people farm. I avoid overstocking and believe in saving plenty of silage and fodder. I enjoy producing quality stock and have never minded the extra work and expense that it entails. This emphasis on detail often frustrates my family but the thrill of landing a top sale price or winning a championship at a major show quickly changes their mood.

In 2002, Mentor Books in Dublin published a set of 4 short books for me called the 'Key Readers'. They included 'It Really Hurts', 'High Ball', 'Grounded' and 'Mr. Fox at the Dublin Horse Show'.

On January 7, 2003, the day of the IFA Tractorcade to Dublin to protest at the falling meat prices, my world fell apart when I was diagnosed with bowel cancer. Immediate surgery and hospitalisation was followed by chemotherapy and radiation. I was forced to stay at home for months and could only hear from family and friends how my animals were doing on the farm. All of the sheep had fortunately been sold the previous September but still there were over forty cows to calve and two mares to foal.

I decided to give up my teaching practice and concentrate on getting strong enough to farm and write again. Now two and a half years later, I am still coping with the unpleasant side effects of the cancer treatment. I am still not as physically and mentally strong as I used to be, but with Gerry's help my interest in cattle and horses remains firm and I have started to write again. We have won prizes in the show sales for my weanlings in both 2004 and 2005. Similarly, my young horses and ridden horses have performed well in the last two years, winning championships and red rosettes in Dublin, Cork, Limerick, Tullamore and Galway. My daughter, Amy, enjoys accompanying me to the horse shows and has had tremendous success both as my handler and rider. She will start an Equine Management degree in September 2005 in Enniskillen. My son, William, on the other hand, enjoys the thrill of piloting racehorses at weekends with his Dad, whist he is in his last year of secondary school.

In summary, two favourite poems come to mind to describe my life. The first, 'The Pasture' by Robert Frost speaks of my youth and fascination with nature:

> I'm going out to clean the pasture spring;
> I'll only stop to rake the leaves away
> (And wait to watch the water clear, I may);
> I shan't be gone long – You come too.
> I'm going out to fetch the little calf
> That's standing by the mother. It's so young,
> It totters when she licks it with her tongue.
> I shan't be gone long – You come too.

Secondly, 'An Irish Blessing' by an unknown author speaks of my love for land and faith in the Almighty:

> May the road rise to meet you
> May the wind be always at your back
> May the sun shine warm on your face
> And rain fall softly upon your field
> And until we meet again
> May God hold you in the palm of His hand.

And finally, if you asked me today, where you are likely to find me tomorrow, it would be, 'down at the farm'"!

Elizabeth Ormiston

Elizabeth Ormiston is a something of a stoic. She works incredibly hard, and I have always had the feeling that she is her own harshest critic. Her fortitude in the face of adversity has always been an inspiration to me, and she is one of the women I think of if I am ever tempted to feel sorry for myself.

If she had been born a few months earlier, Elizabeth would have been eligible for US citizenship. But shortly before her birth in August 1961, Elizabeth's family had relocated from America to Ireland. Elizabeth's father, Peter Olwill, was the second son of a north County Meath farmer. Like so many young men of his generation, Peter had emigrated to New York in 1952, and got a job with an oil company.

It was while he was in America that Peter met Veronica Downey, a farmer's daughter from County Leitrim. When they married in 1960, they decided that they wanted to raise their family in Ireland. While he was still working for the oil company, Peter purchased his first farm in Shancarnan, Moynalty, and it was here that Elizabeth was born shortly after the family moved to Ireland.

Peter subsequently purchased two more farms. He and Veronica worked hard, raising their family of four – after Elizabeth came two more girls and a boy. Peter and Veronica, who have since retired, still live on the original farm. Under the age-old laws of Irish inheritance, it was Elizabeth's young brother who eventually inherited the complete farming enterprise.

"When I was growing up on my parents' farm, it was the mixed farming enterprise that was common in that era. There was everything from dairy cows and sucklers to sheep, pigs, ducks, hens, turkeys and chickens. There was also a small amount of tillage, including potatoes, oats and barley. Being so involved in farm tasks was second nature to me from a young age. As I grew up, I developed a deep love and interest in farming."

Elizabeth's father had two brothers, each of whom farmed similar-sized holdings. The three brothers did a lot of their work together, and were thus able to make considerable labour and purchasing savings. They saved on the cost of hay, picked potatoes together, tied their oats, and all helped one another perform the labour-intensive tasks. They purchased and shared machinery and equipment between them, and managed to work all their lives in a rare display of harmony.

"I always thought of this like the 'meitheal' (co-operative) of old. One of the benefits of the three brothers getting on so well is that they have always been there for each other and each other's families. To this day, although they are advanced in age, my two uncles are always ready to lend a hand, whether it's helping with herd tests, or moving the stock."

Like many other farming families, Elizabeth's family was largely self-sufficient, and met all their own meat and milk needs. "I still remember watching my mother churning the milk to make the butter. We grew all our own vegetables, and the only things we needed to buy outside were commodities such as tea, sugar and flour. The animals were fed by crops grown on the farm, occasionally augmented by feedstuffs purchased from the local co-operative society."

Elizabeth attended primary school in Moynalty. She found her studies interesting, and enjoyed a curriculum that included a wide variety of topics, not all of them exam-related. "The teachers stimulated us to develop an interest in social and rural affairs."

After completing primary school, Elizabeth spent five years studying at Eureka, Convent of Mercy secondary school in Kells. One of Elizabeth's best friends at school was Maureen Ormiston, who lived in the neighbouring parish just over the border in County Cavan.

"On one of my visits to Maureen's home, I met her brother Philip. The attraction was mutual, but because I was only 14, we were not allowed to 'step out.' By the time I successfully sat my Leaving Certificate and completed a book-keeping course, Philip, who was a few years older, had already settled down in a house and small farm that he bought outside Mullagh. He was employed on shift work in the laboratory of a local factory. He was greatly attached to the land, and would spend his days off working as a farm contractor."

Elizabeth's first job was as a book-keeper in the office of an agricultural merchant in Kells. She and Philip married in July 1981, and their first son, Peter, was born in August 1982. A year later, second son P.J. arrived, and at this stage of their life together, Elizabeth and Philip agreed to swap holdings with Philip's parents, who wished to retire from dairying. In June 1985, Elizabeth and family moved to the other farm, and daughter Rosemarie arrived a month later.

"My mother-in-law is the best and hardest working farm woman that ever was, and it was she who taught me how to do the milking. The move to the other farm marked a major change in my life. I was now in charge of running the family farm on a day-to-day basis. I had to see to the milking of the cows, the feeding of the calves, and a host of other farm chores. I also used my training as a bookkeeper to look after the business end of things."

Philip worked twelve-hour shifts and also did contract work for other farmers. Whenever he was available, he would help Elizabeth with the milking. Philip and Elizabeth extended their holding to allow them to keep their calves and sell them for further fattening. In March 1988, they purchased an adjoining holding, and created the property that is still farmed to this day. A second daughter, Eleanor, was born in 1991, followed in 1993 by the baby of the family, Ruth.

Philip loved farming, and took after his mother in his deep attachment to the farm. As a progressive young farmer with over forty acres, Philip needed all his energy.

"He was forever on the move, looking for ways to develop and improve his property. He set about rock-breaking, reseeding and fencing the newly-acquired property. He then embarked on some major development work. He built a milking parlour by himself, he converted a cubicle house to calving pens, he built slatted sheds, and he added a new silage yard. While from a financial and managerial point of view, these were difficult years for us, they were also the best, the most vibrant and the happiest days – if only I had realised it at the time."

Elizabeth's world fell apart one day in August 1995, when Philip was diagnosed with a brain tumour at the age of 38. The doctors only gave him months to live, and he left his job in early 1996.

"After Philip's illness was diagnosed, I had to go back to work in order to help the family cope financially. I revised my old skills, and handled the accounts in a general merchant's on a part-time basis for four years. My flexible hours allowed me to be around to look after Philip. His state of health oscillated between bearable and unbearable, but he held on for far longer than the doctors had said."

Philip eventually succumbed to the tumour in January 2002, and died at the age of 45. His funeral was the biggest ever seen in Mullagh. Elizabeth now worked full-time on the farm, and Peter and P.J. helped her with the dairying for several months. And then, in October 2002, less than a year after Philip died, the cows tested positive for Tuberculosis (TB).

"I tried to keep a brave face in front of the children, but I will never forget how the colour drained from Peter's face when they got the news. That night, I cried my heart out, but I got up the next day determined to move on. The children and I were devastated by the trauma of watching the cows being taken away in two lorries. Luckily, my two generous-spirited uncles came and stayed around for the whole of that awful day."

The valuer's decision, that there would be no more dairying on the farm, came as a hard blow. Milking had been such an integral part of the lives of Philip's parents and grandparents. After much deliberation and soul-searching, and after receiving a clear herd test in July 2003, Elizabeth decided that she would continue farming.

This time, she opted for a suckler herd, and purchased commercial and pedigree Simmental in-calf heifers. They are docile, they make great mothers, and Elizabeth dotes on them.

"The boys have both followed in our footsteps. They both took an active part in farm work, and each in their own way developed a great love of farming. Peter is particularly interested in the livestock side of farming, and has his own pedigree Limousin cows, a flock of sheep and a few horses. P.J. is more interested in the machinery and maintenance side."

After completing their Leaving Certificates, both boys chose to do their apprenticeship in refrigeration. Peter has already qualified, while P.J. has a year and a half to go before he finishes. P.J. came tantalisingly close to winning the refrigeration class in the National Skills Final Competition in 2005, just one point behind the winner.

Rosemarie is studying Humanities at St. Patrick's College, Carlow, and Elizabeth is convinced that she will make a gifted primary school teacher. Eleanor is in second year in Eureka, Convent of Mercy secondary school in Kells, where Elizabeth herself studied. Eleanor has a great love for animals, and is often to be found outside with the livestock. She is also big into art and football, and plays on the County Cavan ladies' football team. Ruth is in fifth class. She is a very outgoing and happy girl, with a great love of sport, and a willingness to help out on the farm.

After Philip was diagnosed with his illness and had to stop working, he had more time on his hands to attend IFA meetings. "Our 15-year-old son Peter used to accompany Philip, and one evening they arrived back home with the announcement that young Peter had been elected secretary of the Mullagh branch. He didn't really like it much, so guess who used to help out behind the scenes!"

Eventually, shortly before Philip died, Elizabeth took over from Peter as secretary. In 2002, Elizabeth sought election as Animal Health representative, believing that her experience with the TB outbreak equipped her perfectly for the post. However, she gave in to pressure to yield to another candidate, and instead she was elected Public Relations Officer of the County Cavan Executive.

Farming, says Elizabeth, is a singular lifestyle. "The IFA has played an important role in providing me with a source of knowledge, and in keeping me up-to-date on current developments in the farming world. The IFA also provides me with a social outlet and a way of getting over the loneliness. I am constantly attending various meetings, and I love meeting so many different interesting people. The IFA gives people an opportunity to voice personal opinions on relevant topics, to highlight problem areas, and to listen to and help solve other people's problems. I think that more farm women should be encouraged to become involved in professional farming organisations. It's very sad that women have never received the recognition they deserve for their involvement in Irish agriculture. For years, women have been the unpaid labour and financial controllers on farms, yet they have had no input into policy-making."

Elizabeth's other activities include membership of the Virginia Show Committee, she is Assistant Secretary of the North Eastern Simmental Society, and she is involved in other rural non-farming committees and associations. "I would like to see women who are already active, encourage others to participate - even though I have to admit that this does not always happen. I sometimes despair that women can be their own worst enemy in this respect."

In order to rectify this, Elizabeth volunteered to become a member of a core group that helped organise the network of IFA women representatives. Elizabeth is hopeful that the new Irish Minister for Agriculture, Mary Coughlan, will encourage greater female participation in farming organisations.

"I must say that I quite enjoy being my own boss. I thoroughly enjoy my life as a farmer, and I believe that living an outdoor life in a healthy environment is great. Despite the uncertainty that decoupling has produced, I continue to retain a positive attitude to the future of farming in Ireland."

Margaret A Gill

While Margaret A Gill may have the distinction of being the oldest among the farmwomen featured in this book, this sprightly seventy-year-old grandmother is definitely one of the youngest at heart. There is something about her sound and solid presence that appeals to me. She has experienced so much, and is always more than happy to encourage and support others.

Margaret has a very young attitude, and on the trip to Australia, she displayed her positive spark. I love her no nonsense liberal approach to life.

If her mother had listened to the advice of her doctor, Margaret would never have been born. This is because her mother had nearly died giving birth to Margaret's elder brother, and was categorically warned by her doctor that under no circumstances was she to have more children. Margaret's mother decided to ignore the advice of the medical profession, and gave birth to a healthy baby girl, Margaret, in 1935.

Margaret and her family grew up on the farm run by her parents on Merrion Road in Dublin. Today, this area is practically regarded as central Dublin, but in those days it was a distant suburb of the capital city. Margaret's father kept a herd of cows, and he delivered milk throughout the Blackrock area. Margaret still remembers the cows walking along Merrion Road near Ballybrock in County Dublin, as they made their way to and from the farm several times a year.

"In my childhood, motorised vehicles were still a rarity in Ireland. I remember the Guinness horse-drawn carts arriving at the farm, carrying grain to feed the cows. Later, they started making deliveries by motorised trucks."

Margaret's father was the youngest of four brothers, and the family was the only Church of Ireland family in the road. The young Margaret attended Donnybrook C of I school, which sounds like a grand description for what was in fact a one-teacher operation. While the teacher taught one section of the pupils, the other section did class-work. After leaving primary school, Margaret attended the Diocesan Girls School on Adelaide Road, and retains fond memories of the many Jewish friends she made there.

"I was always an outdoor girl, and I loved to go on nature activities with the girl guides. I rode my bike everywhere in Dublin. I just loved cycling, I loved roaming around, and I was always interested in keeping myself physically fit."

After five years in secondary school, Margaret left to get a job. "In those days, most respectable girls had a choice between becoming a teacher or a nurse. I was determined not to become a financial burden on my parents, because I did not think it was right for them to have to keep me. So I completed a short secretarial course and went along for a job interview at the Jacobs Biscuit Manufacturing Co. on Bishop St in Dublin."

Margaret's job interview was memorable. On her way to the office, she fell and cut and bruised herself all over. When she entered the office of Mr Bewley, one of the Quaker bosses of Jacobs, he said to her, "You look like you've been through the wars." But he gave her the job anyway. Margaret received on-the-job training, and joined the personnel department. She loved it.

Every summer from the age of six, Margaret had travelled to her paternal grandmother's home in Clonbullogue, near Edenderry in County Offaly. She spent at least a week on her grandmother's farm, and thoroughly enjoyed her annual visit.

"From the age of eleven, I had an even greater incentive for visiting my family. I looked forward to my summers because I had met William (Bill), who lived near my grandmother's farm. Throughout my teens, through school and work, we courted every summer. By the time I was twenty-one, I was sure that I wanted to marry Bill, even though my mother still thought I was too young."

There was no question of Margaret retaining her job after getting married. Margaret still remembers that she received the princely sum of £4 as her PRSI (Pay Related Social Insurance) refund. As she pondered what to do with this money, her father-in-law suggested that she "put that money under four legs." So Margaret went out and bought a calf, thus becoming a stakeholder in her husband's farm. Margaret's exposure to the Dublin job market had made her more money savvy than many other young ladies of her age, and from the time she got married she was careful to protect her own rights.

"But by the same token, my city farm girl upbringing meant that I had never really done much farm work on my home farm. I had worked outside the farm since leaving school, and I was not a natural farmer when I married Bill. Actually, I got into a bit of a panic, and decided that I did not want to make a fool of myself around the house once I moved to Bill's farm. So I enrolled in Ballsbridge Technical School to learn how to cook, wash clothes, and how to iron."

Bill's mother had died before Margaret and Bill married, so when Margaret arrived at Bill's farm, Bill's father was managing the household. Any gaps in the instruction Margaret had received in Ballsbridge were now filled by Bill's father. He was an excellent housekeeper, and he was only too happy to induct his new daughter-in-law into the secrets of homemaking.

Her crash course in Ballsbridge, combined with personal tutoring from her father-in-law, served Margaret well. Before long, she was keeping the house, making butter and baking like a seasoned veteran. "I also picked up a lot of knowledge about the farm, and I learned to drive a tractor. Originally, the family farm was dairy. Then it was cattle and sheep, and now it's just cattle. From an early stage in my marriage, I learned to grow my own fruit and vegetables, and for many years we were practically self-sufficient."

Over the years, Margaret had five children. She calls them "my pride and joy," and she loves them to bits. Barbara was born 9 months and 1 day after the wedding. Now 47, she is a teacher. Sandra, now 45, is a nurse. Caroline, now 44, is a physiotherapist in Canada. "The girls all loved the farm, but at that time, it would have been unheard of for a girl to take over the farm."

After the first three girls, Margaret took a little break, and then came Ken, now 40, a farmer, and Andrew, now 36, who is an accountant in London. When it was time to consider the children's education, Margaret and Bill naturally chose The Kings Hospital boarding school, which was then exclusively Church of Ireland.

Sending the three oldest girls to a boarding school was a financial burden. But Margaret and Bill made a conscious decision to scrimp and save, and to make financial sacrifices, so that the children could receive a good education that would give them choices in life. Kings Hospital also provided Margaret with a new social set, and she became friendly with several of the parents of other children at the school.

"Socially, I think I was doubly disadvantaged. I was a blow-in, and I was regarded as a city girl. I became a founder member of the Irish Countrywomen's Association (ICA) in Clonbullogue, which helped me become more integrated into the community. The ICA gave me the opportunity to meet other women like myself, and I always took full advantage of the social side of membership. I really enjoyed the ICA's educational activities, and avidly attended lectures and seminars."

Because of her city background, Margaret was always more aware of modern trends than many of her contemporaries. She also had a secondary education, which at that time was relatively rare for rural girls. This helped make Margaret more receptive to new ideas. She knew about the latest amenities, and never subscribed to the "If it was good enough for my parents, it's good enough for me" school of thought. It was at an ICA talk by a home economics advisor that Margaret first heard about the new American invention, a freezer, which was now available for the first time in Ireland.

"I promptly bought a freezer, and became one of the first in the county to own one of these new-fangled modern contraptions. Neighbours, friends and official visitors flocked to see my freezer from far and wide. At a time when phones were still a novelty in many farm households, I persuaded my father-in-law to install a phone."

When Margaret's youngest son started school, and the home was empty of children for the first time in years, Margaret decided that it was now time to do something for herself. "I felt it was now my time. Ever since I was a child, I had been something of an exercise fanatic. My love of cycling had helped me stay fit, and knew that regular exercise was vital for my health. So I reduced my involvement with the ICA, and enrolled to do a keep fit correspondence course with the Swedish School of Physical Therapy in Dublin. I used to travel to Dublin at weekends for the practical work, and after a couple of years, I obtained my diploma."

Margaret was now ready to put her training into practice, and to earn a bit of money that would help defray the cost of the children's education. In order to advertise her keep fit classes, Margaret placed small ads in the local press, and she put up notices in church halls and social centres. Farm women started signing up for the course of 6 sessions. Margaret reckons that over the next few years, practically every farm woman in the county came to her classes.

"Many of these women had never before left their homes for a social or recreational activity. Some of the women were so unfamiliar with keep fit that in the beginning, they would arrive in a skirt. The keep fit classes were a lot of fun, both for me and for the women. The atmosphere was very friendly, and it was a great way of socialising."

Margaret stopped her classes with the advent of aerobics. She had never trained with music, and felt that she did not want to change. After giving up the classes, Margaret decided that it was time for further study, and she did a Diploma in Psychology and Assertiveness, "one of the best things I ever did."

When Margaret's father-in-law became ill, it was Margaret who nursed him during the last year of his life in 1975. Two years later, Margaret's own mother died. Her own grieving for these two close relations was made more difficult as it also signified the loss of two beloved grandparents for Margaret's children.

The departure of Margaret's youngest son for boarding school was very traumatic. It meant the last chick leaving the nest, and Margaret decided to take up swimming and walking.

Margaret has distinct views on the rural-urban divide. "I still have vivid memories of the big 1966 farmers' rights march that took place from Bantry to Dublin, followed by a major rally of up to 30,000 farmers outside Government Buildings. I believe that this was a watershed in the relations between Ireland's rural and urban populations. Prior to the march, the divide was very marked, and city dwellers had very negative attitudes to the farmers. During the rally, I remember Dublin housewives bringing apple tarts and sandwiches out to the demonstrators."

It was traditionally the eldest son who inherited the farm, but Margaret and Bill always believed that this would work only if the son also wanted to spend the rest of his life on that farm. "There was an understanding in return for inheriting the farm, Ken would care for us after we retire. At the same time, we would make sure that the other children received the education they required. I think this was a fair system of give and take, with everything out in the open and no room for unpleasant surprises."

After leaving The Kings Hospital, Ken attended agricultural college, and it was expected that he would seamlessly move into managing the farm with his parents. "But then Ken decided to take a year out in New Zealand and Australia, leaving Bill and me to do everything on the farm. For a few anxious months, we were worried that he might not return, but he did."

Ken worked with his parents, taking over the paperwork from Bill, and gradually taking over the management of the farm. At the recommendation of Margaret's accountant, a three-way agreement was drawn up whereby Margaret, Bill and Ken each had an equal share of the farm.

Because Margaret was named as a partner, today she has a full pension, unlike many other farm women. "This is an issue that I feel very strongly about. I don't think it's right that farm organisations are not doing more on behalf of farm-women about such important issues. I also don't think they are doing enough about farmers who are entitled to only half-pensions."

Katherine O'Leary

Katherine is an amazingly strong woman who I can only admire for how she deals with adversity. I find her very grounded in everything she does. She is involved in so many different activities and organisations, especially disability rights. Despite the knocks, Katherine always emerges with a smile.

I will let her tell her story in her own words:

"At this moment in time, in my life, my story begins where it ends with my mother and dear friend. Her death, unexpectedly and untimely has left a void that is indescribable to anyone who has not yet lost their mother, there is a constant sensation in my chest, which feels like there are steel claws clutching my heart, almost preventing it to beat.

I understand this to be the pain of grief. This mother lived her life through us, about us and for us, unconditional steadfast love. Our pain was her pain. Our joy, her happiness and our success, her hope. I was not ready to lose her for many years to come. So much emotion, so much left unsaid, undone. Christmas 2004 and New Year 2005 have been just lonely. I am going through the motions of everyday life supported by the love of my family, I do know that I will get better, but I don't imagine that I will ever be the same again.

I believe that every event in my life has moulded me in some way; the person I am today is the product of my life's experience. I am continually growing and changing, responding to the challenges that family life, and farm life brings to me.

I was born on 22nd of May, 1959, the first child of Maria O'Flynn from Castlelyons, County Cork and Phil Bowe from Donoughmore, Johnstown, County Kilkenny. Daddy died from an aneurism when I was twenty one months old. Yes, I would love to have known him. Mam always said that I was like him, that I had the same ability to make people laugh, and the same determination when it came to work. He loved horses.

When Daddy died, Mam was just pregnant with my brother Phil. My memories of those early years are full of tension and fear. I remember clearly, Mam, Phil and I sitting on the floor at the bottom of the hall of our house, a typical bungalow, all doors closed, holding hands, waiting out a thunder storm. I was scared and in hindsight I know Mam was too! When I was seven years old, Mam remarried Daddy's first cousin, John Campion from Crosspatrick, County Kilkenny. Life began again for all of us. Mam was so happy, Conor and Bernadine were born and our family was whole, busy and content. Sometimes we called John 'The Boss' and over the years this was the easy name we used, not to be confused with the Daddy that Phil and I never knew.

Some memories are more vivid than others. Tears flowing, I shout, 'I will not do it, I cannot do it, I hate it, everyone will laugh at me'! The dreaded parish talent competition had come again, my mother is the teacher. I am 11 years old, standing against the wall in the kitchen between the two doors. I have to recite 'The Mother' by Pádraig Pearse, I am distraught but I cannot escape, there is no escaping Mam. The line 'we suffer in their coming and their going' Oh dear God, how could I explain to her, how the boys laughed and repeated this line, to stand on stage and refer to childbirth! Horror of horrors, I was truly mortified! Worse even, I am dressed in patriotic green, white and gold crimpline and yes I go on to win. Mam is so proud and despite all the heartbreak, I probably taste my first experience of feeling confident.

I recall that time now as I often do because I believe that my mother instilled in me the power to conquer, to face whatever life threw at me and to get up and start again, just as she did.

The little village of Moyne, in County Tipperary where I was born and reared, with Phil, Conor and my sister Ben, is very dear to my heart. It is a place typical of rural Ireland, farms of varying sizes, surrounded by the bog, where everybody knows you, and everything about you. This is both annoying and comforting, but especially comforting in times of sorrow. We lived on a small farm and Mam was the local national school principal.

I loved farming, just as far back as I can remember. Phil and I often milked a cow by hand before we went to school. I loved the smell of the cow and the heat of her where I pushed my head against her so she knew exactly where I was. Dad was a great singer and it was in this stall we learned to sing songs like, 'The Rose of Mooncoin', 'Boolavogue' and 'The Judge Said Stand Up Lad'.....As time passed, we got a milking machine, more cows, and that way of life changed. My farmyard chores changed to feeding the pigs. We had about a hundred of them. Pig sale day in Roscrea meant we got cream buns.

Following my time in national school, I went to boarding school for five years to the Mercy Convent in Cahir, my grand aunt Lizzy was a sister in the convent, so that was pre-ordained. I missed home but learned to stand firmly on my own two feet. As soon as I was sixteen, I became a member of Moyne/Templetouhy, Macra na Feirme. We had great fun travelling around to field days and various activities. The local lads looked after us well and we were never short of a safe lift. The transition to the third level of education was easy for me, having been in boarding school.

I spent three years studying to become a Farm Home Advisor at the Munster Institute (MI) in Cork. I loved every minute of it. There I was to be mentored by one of the most formidable women of her time. Ms Maura E Fennelly, affectionately known as Ma. She instilled in me skills that I use every day. I joined Ballincollig Macra branch. I had a wonderful time and made friendships to last forever, the most important one being that of my husband Tim.

Tim was a local dairy farmer, and besides being gorgeous, the farming bit pleased me also. I fell in love easily; it took Tim a bit of time! He had a few oats to sow first! I qualified in 1979, worked as a farm home advisor in County Laois, I loved the work, but missed Tim and my friends in Cork desperately. As soon as a teaching position became available back at the M.I., I went for an interview in Agriculture House in Dublin, landed the job, and was back in Cork. At this time both Tim and I were prominent members of Macra in County Cork. We did a lot of debating which has been of great benefit to us both in our various activist roles. Macra at that time was vibrant, educational, and great fun.

I became Miss Macra in 1981; it was truly a great time. We were carefree and happy. We made many friends up and down the country through Macra and many of them are now our IFA friends. We were married in July 1982, both of us 23 years old. Our new home and the farm were our main priorities. I gave up work in 1983 to farm with Tim. It was a special time.

On the 3rd of March 1985, though not due until the 20th May, our beautiful daughter Julie was born weighing 3lbs. We were so full of hope and so naive. Doctors talked about developmental problems, but we didn't hear or listen. Nobody really explained. There followed nine weeks of neo natal care. We became very friendly with the nurses, as one does, especially one special lady called Joan Cashman. Joan encouraged us to cuddle Julie; she was so tiny that she could be wrapped in a nappy.

Julie came home in May, time passed. In my heart I knew things were not right. Julie was not sitting, and dragged her legs when crawling. At 18 months she was diagnosed with cerebral palsy, spastic diplegia. Yes, just like that! I came home to the field where Tim was cutting barley to tell him his darling Julie was – I'm not sure what I told him or even what I thought myself. It is somewhat of a daze, and yet I can almost feel her little body against me as I write, the memory is so vivid. Sun shining, the smell of fresh stubble, the pain, the fear for her was excruciating. It would be some time before I would come to trust in her strength, but I will never lose the pain of her disability, or that feeling of 'what did I do wrong? We picked ourselves up and every day with Julie was, and is, a joy. Tim's parents were next door and both a great support to us.

Three years later, on 26th of September, 1988, our son Diarmuid was born. A beautiful morning dawned, but with earth shattering consequences for Tim and I. The labour ward fell silent, the baby was removed quickly, the feeling of panic, we had been here before. It could not be happening again but it did. Diarmuid was born with a hole in his heart, bladder extrophy and Downs syndrome. We were devastated. We had come to terms in so far as one can with Julie's physical disability, now we were challenged with an intellectual disability.

I was overwhelmed with the unfairness of it all. How could God allow this to happen? This child was so sick. I willed him to die, then I was so consumed with guilt, I willed him to live. It is not possible to find words to describe what that time was like. We survived by leaning on each other, and my mother was always there, helping us, crying for us, praying and encouraging us to pray.

Like Julie, Diarmuid was a fighter and he had no intention of dying. He remained in Our Lady's Hospital for sick children in Crumlin, until Christmas Eve, having come through major surgery. He came home with a catheter in his tummy that we had to fill every night to make his bladder grow; it was only 10% of normal size. He was on medication for his heart which terrified me as the doctor said if I gave him too much I would kill him! He was unable to digest milk so was fed on a strange half digested, foul smelling substance.

Despite all the obstacles, we continued to farm successfully and rear our little family. Our lives were filled with physiotherapy for Julie, and several trips a year to Crumlin with Diarmuid. Tim and Julie spent a lot of time together while I was in Crumlin with D. Hence, their relationship is extra special. Cows provided us with a steady income, but it also meant that Tim could not get away easily.

I am almost embarrassed to admit that eleven months later, on 18th of August, our third child Philip was born. Definitely unplanned, against doctors advice, but what a blessing! He was followed seventeen months later by Colm, on the 4th of January, 1991. He was half planned! We now had four children two of them needing extra special care. It was a difficult time but also wonderful, we were young and had plenty of energy. Slowly we began to realise that children with special needs did not have access to many of the therapies they needed.

The fight that began went on and on. Julie was only receiving occasional physio. Mainly we did it ourselves. We got together with a number of families and employed a Conductor from Hungary, who specialised in Conductive Education which was an exercise regime to help children with C.P. I kept this up for eight years, taking Julie from school, driving ten miles and following the programme for two hours, three times a week. Once she went to secondary school, she could not afford the time.

In hindsight I don't know if it helped Julie or not, I do know we stole a bit of her childhood. She also made some friends that she has kept to this day. These friends are hugely important because she can really share her innermost feelings with them. I remember a traumatic day for her when she was about twelve. I was being the sympathetic Mum, saying 'I know how you feel', Julie turned to me and said, 'Mum, you don't, you don't have cerebral palsy'! This was so true. There is only so much I can do, and that is always hard for me. A mother's job is to make things better and I simply can't.

Diarmuid also needed physiotherapy, speech therapy and special stimulation apart from his medical problems. Speech therapy was and still is the big one. For nine years of his life he had none, for the rest a skeleton service, so unjust.

I was always concerned that Philip and Colm were losing out because Julie and D needed so much time. Philip and D were like twins, for a few years D's development mirrored Philips, and then when Philip moved on Colm was there to help him. Julie read with him and played with him. The three boys learned early on that Julie needed her bags carried and various chores done by them.

When Diarmuid started school, I joined the Parents and Friends committee. Before long I was elected to the Board of Management, and served nine years, my last as chair person. As a family we continually encountered problems, lack of service, waiting lists, and problems with access for Julie. The frustration was never ending.

I responded to this by becoming involved in the fight for better services for people with disabilities. I am an active member of the Council and Executive of NAMHI, the association working for people with intellectual disability. Lobbying for the provision of an effective disability bill is a priority issue at present. The struggle for services for Julie and D has made me strong, confident and capable of speaking in public. I am also a member of the Research Advisory Committee of the National Disability Authority. Sometimes I have to question all my involvements and wonder what am I achieving? Unfortunately, many of my fights with the system will not and have not helped my children, but the ones that have been born since. It is hard to find a balance.

My children's needs have taken me places I would never have gone, I believe that until you have reached the depths of despair, only then can you appreciate the heights of joy. As a family we have learned to value small progress and we celebrate every achievement.

Our children make us so proud. Julie is now studying Applied Psychology in UCC. She is articulate, intelligent and beautiful. She hides her inner struggle well, but as her mother, I perceive her pain and exhaustion as she battles physically to get around and not allow her disability to interfere with her progress. She is a columnist in the 'Evening Echo' every Wednesday. Diarmuid is a young gentleman attending special school, he is confident and trustworthy within his own level of ability, he helps Tim milking each evening. I worry about his future. Philip is in third year, facing his junior certificate, he is also confident, extremely caring, a good farmer and cook. Colm is in second year, super confident, cautious and loving, an avid reader and good farmer. Both lads love to play hurling. We all love a good match and attend as many as we can.

I am chair person of the Parents and Friends committee of their school, Coláiste Choilm. I believe it is important to be involved in the schools activities to demonstrate that I care about their education.

While the children were small, I kept in touch with my teaching, giving adult education classes for the VEC (Vocational Education Committee), in Cookery and Home Management, to low income groups around the city, two to three mornings a week. There were many times that I was humbled by these women who raised families on a fraction of what I had, sometimes suffering abuse.

In 2002 a position as teacher of Home Economics became vacant at Diarmuid's school, in Ballincollig. It was ideal for me as it is near Coláiste Choilm so the lads can travel with me. I work four days a week having Wednesday off to do my other things. Sometimes it is hard to fit it all in, but farming has changed, incomes have dropped, so the extra income is necessary. I like the work, but I do miss the farming, the views at Woodside, and being able to spend time in my garden.

In the last number of years, we have also developed a farm tour business, where groups of foreign visitors come to our farm. Tim talks to them about Ireland, farming and agri politics. We then bring them to the house for tea and scones. I always have the scones made that morning and served with homemade jam from the fruits of our garden. If the boys are around they get ready the living room and do the serving. They have learned great social skills as a result and are starting to use their French and German.

The groups vary in size from twenty to fifty. They love to see how an Irish family lives, and if we're lucky the boys or Julie will sing for them. The tours run from March to October. I always enjoy bringing visitors into our home and we have had many foreign visitors over the years. Many of them are as a result of Tim being a Nuffield Scholar.

Parallel to family life, I continued my involvement with the farm and agri politics. If Tim was going to a meeting or conference, I went too. This kept me in touch with what was happening in the wider agricultural world and meant that Tim and I could have informed healthy discussions, and a few arguments thrown in. We were both involved in the IFA, Tim became county chairman which meant he was away from home quite a bit, and as a result I had to take on the responsibility of the farm during these times.

I found it challenging and rewarding, and this was when I really learned about the farm. He was in Australia and New Zealand for five weeks doing part of his Nuffield Scholarship in 1998. By the time he returned I was nearly ready for my cap and gown in farm management! We built a new house in 1995 and a new farmyard in 1998. We are progressive farmers, and Tim is an astute businessman. We have almost doubled our milk quota through hard work and leases. We are positive about a future in farming for us and for our sons, if that is their choice.

I have been secretary of Cork Liquid Milk producers since 1983! I have worked with a changing committee of twenty men, one woman for a while, and have always been treated as an equal and with respect. I believe women can be their own worst enemies by not getting involved. If you want something badly enough you have just got to go for it. Of course there are barriers, obvious and hidden, but if we don't break them down they will always remain. The representation of women within the IFA is a mere 6% and 13% in the Dáil. I take these figures very seriously, and I believe it is important that each of us works to change these figures by supporting each other.

The Equality Initiative headed up by Mary Carroll for the IFA was indeed a positive measure in the lives of many farmwomen. I was one of those women who had battled along side the men and won my own right of recognition and place within the IFA. It is hard at first to be the only woman, but once you get over that aspect the men get over it fast enough. I doubt if I could have done this without enduring my own personal battles. Strength of character is built only by experience. The Equality Initiative gave a sense of pride and recognition, and opened up new opportunities. It made it easier to stand up and be counted, it was also recognised by the men because it was their own organisation.

In 2003/2004 I won a place on the Power Partnership programme. This was directly as a result of the IFA initiative. The title of the course was 'A feminist view of Politics, the Economy and the State'. It was divided into six modules of three days each at wonderful places like Maynooth and Marino College. The assignments were daunting, it was so long since I had done an academic essay but I passed them and will graduate in March 2005. This was something that I did for me, and it certainly was a new venture for all of us. Tim was sceptical because of the feminist thing.

It was residential, so I had to accept that, yes, the family could cope without me for those few days and also that fathers are just as good at mothering, when necessary, just as we are at fathering! There are many days when I feel the weight of the world on my shoulders, with so much to do, a mothers work is never done. But did I tell anyone that I was under pressure, stressed, feeling unable to cope, or even asked for help. The truth is no.

Last year a group of women, many of us involved with the IFA initiative decided to go on a trip to Australia. This was a chance of a lifetime, and Tim and Julie encouraged me to go for it. I was apprehensive and guilty. Julie would have just begun college and I wouldn't be there to support her. Anyway "Women in Agriculture" became a reality, and thirty-two brilliant women found themselves down under.

We shared our trials and tribulations, we pooled our stories and gained strength from each other. Women had lost husbands, sons and daughters. Serious illnesses had been overcome, disabilities encountered and yet we were there, sharing and caring. A dimension of the trip was to spend five days with an Australian farm family. I stayed with Frank and Marianne Templeton and their three girls. We made an instant connection and a friendship that will last a lifetime. I milked cows with them, learned as much as I could, and was given a great time. I watched the penguin run on Philip Island when the penguins went home to bed from the sea. I swam with the fish on the Great Barrier Reef. I stayed an extra week to renew friendship with my first cousins. It was lovely. But it was just great to come back home. Life had gone on just fine without me, which means I can go again!

After our return, my mobile phone did not ring for two weeks. It was a bit strange. I like to be needed, and most of all I like to make a difference. That fact complicates my life but when NAMHI asks me to do a speech, or the NDA, or the Parents and Friends, or the IFA on behalf of women or people with disabilities, I will always say yes because that is who I am.

All my children are differently abled, wonderful, unique people. I am so proud of them. I have learned to live in the now and not to think too far ahead, otherwise I wouldn't sleep worrying about where D is going to be, because there is no service once he is twenty one. That is the new challenge to be addressed. The answer may be in the Disability Bill if we can make it properly rights based. Who knows?

For some years I have worked during the summer months on the Féile Bia programme, encouraging hotels and restaurants to use Irish produce. This initiative is supported by the IFA, the Hotel Federation, the Restaurants Association, and run by Bord Bia. This is part of our future in farming to secure our markets. In September, I was appointed to the board of An Bord Bia, I am proud and delighted, and will be a worthy member of the Board into the future. Imagine me on a State Board!

There are times when I hit bottom, who doesn't? I am sustained through these times by my many friends. There are a few really special ones and a phone call out of the blue from one of these is enough to lift my heart. I love music and it allows me to escape when necessary. Many songs trigger memories and help me to remember, and almost relive, special times.

On the 16th of December, 2004, we buried Mam, the wind blew and the rain lashed us. The end of an era, the closing of a chapter in my life, the opening of another. Who knows what's around the corner? Ar Dheis Dé go raibh a Anam (May she be at God's right hand).

Frances Coffey

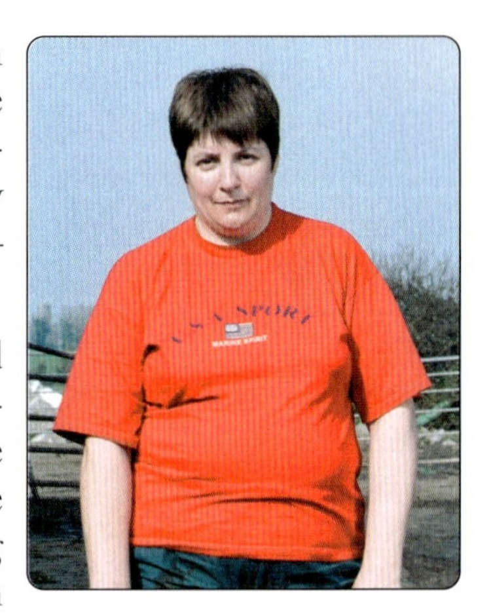

Frances is one of the rocks I have learned to rely on for advice and feedback. She has a powerful positive nature, and she is particularly good at seeing the bigger picture. Her family life makes me positively dizzy, yet all those children don't seem to faze her – she has them all organised like clockwork.

Whoever imagines that a Macra na Feirme field evening in the mart in Cashel is not a romantic setting, never heard of Frances Coffey. For it was here that Frances met her future husband Pat. At the time that Frances joined Macra, a few years after leaving school, it was a very strong and vibrant organisation for young farmers. After Frances met Pat, the two of them started travelling around to various Macra field evenings all over the country. Many of the young people at these meetings have since become active in the IFA.

Frances grew up on a dairy farm outside the town of Templemore in County Tipperary. She was one of nine children born to Mary and John Maher. Frances was a child who loved the outdoors, and she also enjoyed helping out on her parents' farm.

"Pat and I got married while we were still in our early twenties, which allowed us to develop along with our children. Becoming a mother for the first time, with the birth of my first child, Michelle, gave me a new appreciation of my own mother. I still remember my father saying to me: 'From the day this baby comes into this world, until the day you go out of this world, you are responsible.' Over the next few years, I had seven more children, and Pat and I have always encouraged the children to become involved in extra curricular activities. I agreed to take on the post of Assistant Treasurer of CBS Parent's Council in Nenagh because I believe that it is important to be active with the schools where your children attend."

When Frances lost her father in 1991, she had difficulty accepting that he was gone. One day as she attended mass in Dublin, she had a spiritual experience. During the consecration at mass, she felt a tremendous joy welling up inside her.

"Tears were flowing down my face, and I felt a presence – a sort of tingling sensation - on my right shoulder. My aunt and uncle witnessed what happened, and told me afterwards that my face lit up as if a bulb had been inserted in my neck. I believe that at that moment, I was able to let my father go to God, and I am very conscious that his spirit is all around me as he guides me throughout my life."

Frances and Pat practice mixed farming, with Pat working off the farm as manager of the South Midlands Farm Relief Services. The farm is run by Eddie Boyle and David Walton, who have become part of the family.

After working for an accountancy company for several years, Frances decided to become a full-time parent after the birth of her third child. Then in 1993, she returned to working off farm as a job-sharing secretary in St. Joseph's College, Newport. Working as a member of a team was a pleasant change from the isolation of home, and Frances remembers this period as an invaluable opportunity to view the education system from the inside out.

"That same year, I accompanied Pat and his sister Bernadette on a trip to Africa to visit Pat's other sister, Dr. Mary Coffey, who was working in a hospital in western Tanzania. The three-week trip was a real eye-opener. We stayed for a week at a holiday resort in Kenya, and a second week on safari, touring Nagorngoro Conservation Area and Serengeti National Park. But the highlight of my trip was during the third and final week when we stayed with the native people in western Tanzania and participated in their culture. Mary's hospital was located in a remote village called Makiunga, where volunteer medical staff provided astonishingly good medical services with very limited facilities. I felt privileged and honoured to be able to experience this other side of Africa. I learned that to truly appreciate the real culture of a country, you must live with its people. I found the women remarkable. Despite the misery, the hardships, and the poverty they had to endure, they were content with life."

Frances returned home with a renewed consciousness of how fortunate she was. The trip also changed her attitude towards material things. In 1996, she became involved in a local fundraising campaign on behalf of two cerebral palsy children, Margaret Corcoran and Mark Ryan. Frances helped raise money for the establishment of trust funds that would help the children with their physical and educational needs. A number of VIP's were enlisted to help promote the Appeal Fund, and the concluding event was attended by Brian Crowley MEP. This project also marked the emergence on the community scene of Frances Coffey, a person in her own right, and no longer just Pat Coffey's wife.

"I had always wanted to broaden my education. When the opportunity arose in September 2003, I enrolled in the HighWay Programme, an induction into third level education at Tipperary Institute in Thurles. This course, in the middle of which I gave birth to child number eight, allowed me to update and acquire new skills with accreditation. My eldest child, Michelle, was doing her Leaving Certificate at the same time, and my daughter Christine was just starting primary school. I think that all my children learned an important life's lesson – that it is never too late to learn, no matter where you are in life."

The course allowed Frances to reconnect to who she was, and to where she was going. The course also helped her move forward in a more focused and effective way. An integral part of the course was personal development, and Frances found herself looking on life differently. She stopped taking things at face value, she honed her critical thinking faculties. She found that she was devoting more time to reflecting on her life's experiences.

"It was while I was working on the business module of the course that I drew up a business plan based on the need for farmers to diversify. Basic sociology was another module that I enjoyed, since it gave me an understanding of how society operates, and introduced me to different theories. I loved the entire study experience, and used the opportunity to update my computer and mathematical skills. The course convinced me that I have the ability to go further if that's what I want to do. It also gave me the power of knowledge, which leads to greater independence, and to a more enriched life."

Frances' IFA career has blossomed in recent years. As chair of the IFA North Tipperary Farm Family Committee, and representative of the Toomevara branch at the County Executive meetings, Frances has become the voice of North Tipperary at National meetings.

"I regard the appointment of Mary Carroll as the IFA National Equality Officer as a milestone event in IFA history. Mary inspired me – as well as many other women in IFA - to explore my potential in a most natural manner. With her constant encouragement and empowerment, Mary proved to be a catalyst and super role model for rural women. Mary is like a door which opens up access routes into a new era in farming. And it was through Mary that I became involved with organising the first ever Rural Woman's Conference in Croke Park in October 2003. This groundbreaking experience allowed me to meet like-minded women who shared an ambition to improve their lot and to empower one another. I felt a great sense of inclusion, and I was emboldened to believe that I had a part to play."

At the Croke Park conference, Frances met visitors from different parts of the world, including the UK and Australia. So when the Australians invited the IFA women to visit Oz, Frances was determined to make the trip. With the active encouragement of all her family, Frances joined 32 Irish rural women on a three-week trip where they learned from, and engaged with, women who are light years ahead of their Irish sisters. Frances was also interested in how the Australian women had set up a national network, and she returned home determined to come to terms with a changing world.

"I believe that all rural women must be encouraged to make their contribution to their homes, their community and their country. I believe that if you conduct an honest dialogue with yourself, you will get a better perspective on how you are doing. It has taken me a long time to become the person I am today. Self-awareness is the key to a life of happiness and fulfilment. And by happiness, I don't mean getting what you want, I mean wanting what you have."

As someone who has enjoyed an intimate perspective on the farming world both before and since joining the IFA, Frances believes that farmers need to reinvent themselves. More farmers will decide to farm part-time and to seek employment, and more and more farmers will go back into education in order to improve their employment options. The move towards farming on a much larger scale is bound to impact on the farm families. By adopting a positive approach to change, farmers will eventually find workable and successful solutions. Frances sees the general public becoming more in tune with the farmers.

"For years, the farming community suffered from consistently bad press. Now things are changing, as I experienced first-hand during the Tractorcade, when I appeared on national TV and was interviewed on local radio. I used the opportunity to highlight the low incomes of farm families, and the differences between the farm gate prices and the money paid out by the consumer. The Tractorcade was an example of men and women standing side by side in order to fight for fairer prices for their produce. I was greatly encouraged by the solidarity displayed by local businesses during the beef blockade outside the meat factories. For me, this was an encouraging sign that the tide of public opinion is turning."

Frances is fiercely proud of her nine children, ranging in ages from 18 years to 2 months. Michelle (18) has started Law and Accounting at University of Limerick. Marie (16) is a sporty, musical lively character with a sharp wit. John (15) is a very relaxed type, works on the farm and is totally dedicated to rugby. Emma-Louise (14) is a multi-talented girl, very kind, considerate and interested and active in all things agriculture. Bernadette (10) is a stylish and observant girl and extremely loyal. William (8) is involved in all sports, and is very academic and pleasant. Christine (6) is the younger version of Michelle, very active and a good gymnast, Joseph (20 months) is very much an active, outdoor boy. Rounding up the nine is Matthew (2 months,) a very happy and good-humoured child who always has a beaming smile.

Frances has a firm sense that when she departs this life, her real legacy will be the part of her that will live on in this world through her family.

"In my world, the most important treasure is the human being. It is vital to remain positive. I am who I am. There are two phrases that I have adopted as my personal mantras. One is by Sarah Ban Breathnach: 'The authentic self is the soul made visible'. I also love an inspiring phrase I heard from Cathy McGowan, the Australian farming expert I met at the Croke Park conference, and who was also my host in Australia: 'Blossom where you grow.'"

Denise O'Sullivan Breen

Denise O'Sullivan Breen is another of the women I have met whose fierce intelligence has shone like a beacon for me. As a farmer and mother of two small babies, she is an example of what can be done. I see Denise as someone who gets her sleeves rolled up and stuck into the work.

Denise grew up as a carefree child, the fourth of five girls, in a family where both parents worked as farmers. Her father also worked as an agricultural contractor, specialising in grain and to a lesser extent, haymaking. "It was only natural that as children, my sisters and I all helped on the farm, feeding the calves and looking after the hens and chickens. On Saturdays, our job was to roll the barley, something I must admit I never liked. When we got older, we all learned to milk the cows."

The first day of each year's harvest was special for Denise, and the roar of her father's combine harvester certainly added to the excitement of the occasion. The mandatory ride on the combine was, of course, a great treat for all the children. Special in a different way was the day her father would sell the bullocks at the mart. Denise and her sisters were not particularly happy to see the bullocks leave, but the £5 each sister received in the evening made all the hard work of rearing the calves worthwhile.

For Denise, the memories of growing up on the farm were good. "It was a healthy, fun way to grow up. At the same time, it was also educational. We learned the laws of supply and demand through our egg selling business, derived from our dozen hens. We also learned skills from building 'cabby' (play) houses, shops, post-offices, banks and the like. It was all part of growing up."

During her formative years, Denise did not show any particular inclination to fol low her parents into the farming profession. While at secondary school, farming did not even appear on Denise's career radar screen. She successfully completed her Leaving Certificate, and was accepted to study in University College Cork. She moved to Cork, and began her university courses.

"When I went off to UCC, my three older sisters were already pursuing their non-farming careers, and my younger sister was still in school. In the middle of my studies, I had a dramatic change of heart. I decided to give up university and to study farming instead. It's not that I didn't like my university course. I did. It's just that I wanted to get into farming."

Denise enrolled in agricultural college in Clonakilty, County Cork, and completed her certificate in farming in 1991.

Denise had excellent training in running the farm on her own from her father, who helped her gain valuable experience in the earlier years. She took over the operation of the family farm in 1995. She grew to love her work as a farmer, and farming remains her major passion to this day.

"As a woman farmer, I have been acutely aware of some of the problems faced by women in agriculture. The most obvious difficulties that women farmers come up against are of a physical nature. There is no shame in admitting that women are generally not as physically strong as our male counterparts. I quickly learned to work around this by looking for every opportunity to substitute machinery for manual work - mechanical horse-power is better than man-power."

Denise systematically introduced minor changes around the farm in order to allow her to operate in a safer and more practical manner. For example, fertiliser had always been delivered in 50kg bags that had to be manually loaded. At Denise's insistence, fertiliser was now delivered in half-tonne bags and loaded with a tractor and loader.

She applied the same principle to silage. Instead of using a four-prong fork to push it closer to head-feed, she did the work by means of a scraper on the back of a tractor. Meal was delivered in bulk to a meal bin situated near the slatted unit, which avoided the need to transport the bags of meal from a shed at the other side of the yard. It was small but significant changes like these that made life easier for Denise, and facilitated the way she was able to carry out the day to day duties on the farm. In short, she became more efficient and more productive.

"For several years, I worked alongside my father, baling hay and straw on contract. I remember times when I would enter a field with the tractor and baler 'in tow', and the first question I was asked was whether my father was on his way. At first, I used to take exception to the underlying assumption that I couldn't possibly be there on my own".

Denise eventually learned to simply say yes when asked, even though she knew full well that her father could be harvesting anything up to 10 miles away. If saying yes and swallowing her pride meant alleviating the concerns of the particular farmer, so be it. She felt secure enough not to have to explain in detail that she not only transported the baler from farm to farm, but that she actually also operated it.

Men were not the only ones to assume that a woman could not operate independently. Women proved that they could be equally ignorant and unintentionally insulting.

"I still remember one woman asking me whether I drive a tractor. This was after I had already been working on the farm for several years. I often wonder why some women still get asked questions that would never be asked of men. Being asked if I know how to drive a tractor is like asking a teacher if she knows how to read. I know that things have moved forward a little – but still not enough."

Denise no longer worries whether her being accepted in her own area as a farmer is because she has proved herself as a real farmer, or because people have simply got used to the fact. And though she no longer bales on hire, she does continue to deliver grain to the local co-op.

One of the most difficult things that Denise experienced when she first came home after giving up college was the loneliness and the isolation. "I was used to having my friends around me, and I had enjoyed the social scene which goes with being a college student. My lifeline came in the guise of Macra na Feirme. Joining Macra was one of the best decisions I ever made. It allowed me to make new friends in my locality, and to develop new skills such as debating, drama and public speaking. "

Macra also provided Denise with the perfect opportunity to meet her husband. She and Michael were members of the same Mourneabbey Macra branch. "We became friends, and as they say, one thing led to another. Michael is not from a farming background, but this has not stopped him taking an active role in my farming career when the time allows."

During the first few years of their married life, Michael worked for engineering firms, but started his own business in 1998 offering a contract draughting service. He works from an office that they built next to their home, which allows Denise and Michael to see each other – and their young boys - during the working day.

"Getting married nine years ago did not dramatically change our lives, but the arrival two years ago of our first son, Oisin, certainly brought changes – for the better. From a practical working point of view, Oisin has to come on the tractor with me. With this in mind we fitted a complete car seat to the tractor for him".

This certainly drew an interesting array of looks from people she encountered while driving along.

Denise considers herself blessed with good friends who she made through school, college, Macra and her IFA involvement. She is currently chair of Mourneabbey IFA, and is also involved in North Cork IFA. She and Michael both love to travel, and she is looking forward to further travel when the boys are a little older.

Having achieved her ambition of studying farming and becoming a farmer in her own right, Denise decided that the time was ripe for a resumption of her interrupted university studies.

"At first I was a night student at UCC, and then I started attending as a day student. After completing my degree in Economics and Sociology, I went on to do a Masters degree in Women's Studies. I made great friends at college, and the whole experience opened my mind to areas I would not have been familiar with. Returning to university has been a source of huge satisfaction for me and for my family, but no matter how many academic qualifications I possess, I still consider myself first and foremost a farmer – and a very proud farmer at that."

Although busier now with the arrival of Fionn in March 2005 to join his brother Oisin, Denise now consider herself someone who leads a very happy and fulfilling life – as a wife, mother, farmer and friend.

Denise remains optimistic about the future of farming in Ireland. Even though she acknowledges that the overall number of full-time farmers will be reduced, Denise believes that there will always be a need to feed the population, and there will always be demand for high quality food. "Irish farmers, both men and women, are eminently capable of delivering the goods", says Denise.

As for what the future holds for her children, Denise is quite sanguine.

"I see no point in worrying today whether my children will follow my example and become farmers. If they decide to adopt farming as a way of life, I'll be delighted. If they don't, I'll still be delighted, so long as they do something they enjoy. I will always encourage my children to find their own path, and I will always teach them that happiness and fulfilment are the most important goals to strive for."

Eileen Redpath

I have always been struck by the feeling that under Eileen's quiet exterior is a strong and dynamic woman who could do anything. She is one of those fiercely independent women who find it hard to ask for help. I will always be appreciative of the way Eileen stood up for me during a particularly bruising meeting, her quiet but steady voice calming the emotive atmosphere.

Eileen was also very reluctant to have her story told in this book, because she did not believe that her life warranted any special attention. I thought differently, and Eileen allowed herself to be persuaded.

The 9th of January, 1967, marks the date that Eileen Redpath became an NFA/IFA activist. She can trace her farming activism directly to that date, because that was the day that farmers blocked bridges as part of the Farmers' Rights Campaign. It was on that day that she first fully appreciated just how much could be achieved by unity of minds, unity of purpose and unity of action.

As Eileen's family farm was situated beside one of the bridges being blocked, her parents took an active part in the campaign. In the ensuing court cases, when many farmers chose jail rather than paying fines, Eileen's father was let off on a technicality, a fact which he deeply regretted.

The farming organisation was strengthened by farmers helping to do the work that couldn't be done by their colleagues who had gone to jail. The main burden fell on the farmwomen, whose courage and determination was a crucial factor. The sight of hundreds of women, including Eileen's mother, demonstrating outside Portlaoise jail, left an indelible impression on Eileen which has remained with her throughout her life.

When the local NFA branch made a presentation to her father in recognition of his contribution during the Farmers' Rights Campaign, it was Eileen who received it on his behalf. And even though Eileen did not join the organisation until several years later, she grasped the importance of having a unified and strong farm organisation.

Eileen's life began on a farm in County Meath. This was a fairly typical mixed farm for its time, with sheep, cattle and pigs. Eileen's parents were a huge influence on her life. "My father had a profound respect for his fellow man, and he instilled this in both his daughters. He was a great believer in the importance of community service, and he taught his daughters the importance of civic duty. He never tried to persuade us to vote for any particular party, but he did try and persuade us to exercise our right to vote in local and national elections."

Eileen's father had a terrific sense of place, and was very interested in local history. Eileen now regrets that she did not imbibe all that he imparted on this subject, but he did influence her to join the Meath Archaelogical and Historical Society. Here, she was lucky enough to have the then County Librarian, Mrs. McGurl, and noted local historian Dr. Beryl Moore, as her mentors.

Eileen's mother was a very practical person. "Although she herself did not drive a car, she made sure that both her daughters learned to drive as soon as it was legal to do so. She engendered courage and determination in me, and loved telling me stories of the great pioneers like Ernest Shackleton and Albert Schweitzer. Gladys Aylwood, 'the small dark woman', particularly appealed to my sense of adventure. My mother was also very interested in drama and poetry, and used to read to her daughters on Sunday evenings."

Eileen's father was very musical, and could play many instruments. Although self-taught, he also had a very fine singing voice. There were always other interests outside farming, and throughout Eileen's life, music and reading have been hugely important. She has always found that at low points in her life, music has been a great healer.

Eileen's formal education was fairly general. She finished secondary school and completed a secretarial course. But just before Eileen sat for her Leaving Certificate, her mother became very ill. This made Eileen realise just how important her mother was in her life. Eileen's father had suffered from severe arthritis from quite an early age, so it had always been Eileen's mother who had carried the main burden of running the farm.

Eileen worked for several years as a medical secretary, but she always combined this with farm work. Then her father died suddenly at a relatively early age. It became necessary for Eileen to take up farming on a full time basis.

The first thing Eileen discovered was that there were large gaps in her farming education. In particular, she knew nothing about the farm machinery. One day when a very patient man was trying to teach Eileen welding on a course in Warrenstown, she somehow picked up the wrong eye shield and was unable to see properly. She was lucky that the whole place did not burn down.

Suddenly having to assume full responsibility for everything that happened on the farm was scary. She quickly learned that the needs of the animals always took priority. There was no one around to blame when anything went wrong. However, Eileen's mother was always in the background, giving support and knowledge that was a great help, and Eileen's sister also helped at weekends and during holidays, and members of the extended family also lent a hand. As difficult as this was, Eileen discovered that experience is a great teacher, albeit a very expensive one.

Like her father before her, Eileen was a member of several community organisations. Some years earlier, at the instigation of members of the local NFA branch, Eileen had helped set up a local branch of Macra na Feirme. This proved to be an important learning curve for Eileen. She learned the value of listening, and she gained a new appreciation of achieving consensus when contentious issues are at stake. Eileen represented her branch on a number of occasions in various competitions at county and national level.

Once farming became her sole occupation, Eileen joined the IFA. She had learned from the Farmers' Rights Campaign how crucial it was for farmers to be part of an organisation representing their interests. She had no idea that she would often end up as the only woman at IFA meetings. In her view of things, the IFA was a place where farmers meet to discuss mutual problems.

Eileen soon found herself being elected to various positions within the organisation. This was curious for her, because earlier in her life, Eileen had refused to enter party politics. And now here she was, embarking on politics of a different type. Eileen's IFA career followed a similar path to others who achieve office. She started out as branch secretary, then representative to the county executive, and then branch chair. There was often canvassing involved, which at first Eileen found quite daunting.

Eileen participated in much footpath slogging in the various farmers' protests, particularly against the 2% levy, the super levy, and the iniquitous Land Tax. She was also part of a delegation to the European Parliament in Strasbourg where she protested about the "cheque in the post" syndrome. Kildare Street in Dublin and the surrounding streets became very familiar over the decades. Eileen remembers being present at the famous lamb episode at the Department of Agriculture, when the sheep did what sheep do – and followed each other into the Department offices.

Eileen was elected to succeed the very able Pearl Baxter, the first Meath county representative on the National Farm Family committee. The contribution of farmwomen to agriculture had originally been given a major boost when then President, T. J. Maher, set up the National Farm Family Committee within IFA. The committee's first chair was Betty Fahy, a truly inspiring farmer who later went on to become County Chair in Kildare. At the time, she was one of the very first women county chairs in IFA. Many years later, Eileen had the privilege of meeting this remarkable woman who had blazed a trail for farmwomen.

When Eileen joined the Farm Family committee, she experienced considerable culture shock. "Having been used to being almost the only woman at a meeting or at a mart, here I was surrounded by women. It took me quite a while to feel comfortable in this new environment. Eventually, I found mutual ground with my colleagues, especially in relation to the effect of stress on farm families. This was at a very difficult period in farming, when dramatically rising interest rates were taking their toll on farmers."

Eileen also represented IFA on the Council for the Status of Women, now the National Women's Council. This was even more of a cultural shock, because there she discovered the mindset that that anywhere outside Dublin was regarded as rural. She did not have an easy time on this body. Several women from other organisations wondered out loud why the IFA deserved a representative on the council.

Eileen says that she entered the Council for the Status of Women with quite liberal views, but by the time her term was up, she was more likely to vote with the so-called conservative block within the Council. But all in all, she met many amazing women who were very inspiring and encouraging.

This was also at time of the first Abortion Referendum, and the council had voted in favour of some limited abortion. Although Eileen refrained from voting on this, because the IFA 's constitution is strictly non-sectarian and non-political, this did not prevent her receiving some very unpleasant phone calls. Eileen remembers the referenda as a source of great divisiveness throughout the country.

Eileen's next position was as IFA County Secretary. By now, she had a greater insight into how the organisation worked, and she learned even more in her new position. But nothing quite prepared her for her role as National Council representative for Meath, which coincided with the Presidency of Meath man Thomas Clinton.

As Chairman of the National Taxation & Credit Committee, now the Business Committee, Clinton had been instrumental in helping very many farm families during the distressing period of rising interest rates. Nevertheless, he incurred the wrath of several key individuals within the IFA by suggesting a number of major structural and procedural changes. Ironically, many of these same changes were recommended years later by the Dowling Report, and passed by National Council.

So Eileen's time on National Council was not a particularly happy one. She did not seek a second term of office, as she was fed up with receiving early morning and late night calls offering advice. She decided that it was time to take more of a back seat in the organisation.

When Eileen was asked once more to represent Meath on the Farm Family Committee, she reluctantly agreed. Her second time round was a lot more effective. She was helped by the fact that by now there was a thriving Farm Family Committee within the county. Eileen felt much more solidarity, and she felt lucky to be part of a bunch of highly energetic and capable women.

A casual remark made one night at a meeting proved to be quite transforming.

Eileen would not describe herself as very social, and she had never considered having tea after meetings. But when someone suggested this idea, she agreed to it – and the whole group became much more cohesive. The teatime became almost as important as the meeting itself, and encouraged more members to come out of themselves and make a valuable contribution to the proceedings.

As a strong exponent of life-long learning, there have been few times in her life when Eileen has not been involved in some type of study. She was raised on the maxim that a person's reach should always exceed their grasp. She has always believed in pushing out the boundaries, both physically and mentally. She spent two years doing a counselling course in Maynooth, which she found extremely helpful because it taught her not just about other people but about herself. The course also taught the value of not judging. Eileen was impressed with the Chinese saying: "You shouldn't judge a man until you have walked the length of a moon in his shoes." She tries to carry this with into the many situations in which she finds herself.

In 2003, Eileen embarked on a Community Studies programme promoted by LEADER in conjunction with the Dundalk Institute of Technology. This comprehensive outreach course of studies gave people the opportunity to participate in third level education. The course provided detailed insight into many different facets of community activity. Eileen loved the freedom of letting her imagination soar, particularly in the communications module.

"I feel strongly that too many people of my generation had their imagination 'surgically removed' at primary school level. The Community Studies programme empowered and invigorated me to unlock the creative imagination that had been locked away since I was five years old."

Eileen has always been concerned that the IFA's effectiveness on the economic front has hidden the organisation's reluctance to examine the many social problems affecting farmers and their families. There are so many causes of stress, including unnecessary red tape, disease problems, the sudden death of a partner, or simply the fact that a great many people farm on their own.

Eileen believes that this has not been sufficiently addressed by the IFA. For example, it took almost 20 years for the Farm Family Committee to produce a leaflet dealing with stress. The IFA is finally starting to show some concern for men's health. Eileen believes that the IFA must do more to provide emotional support to members. Even in the Farm Family Committee in Meath, the counsellors who dealt with a variety of issues were only scratching the surface. As Irish farming enters a new phase following decoupling, new problems and stresses will arise for which we are not prepared.

Eileen knows something about the stresses that can befall farm families, as she had the misfortune of suffering three major stresses within six months. First, her beloved mother died, then her herd suffered a major TB outbreak, resulting in the loss of a third of her animals, and finally she was attacked by a cow that almost dispatched her to the next world. Eileen says that her survival is due to a number of factors, including her good friends and a terrific extended family, who sustained her throughout her ordeal. She was also sustained by a firm and absolute belief that she would eventually come through the tunnel.

In addition, Eileen's years as a farmer have given the kind of grittiness that is the hallmark of so many farm families. She was particularly helped by an Ennegram session run by Teagasc's Celine McAdam. The self-understanding that Eileen took from those sessions was a major motivator in setting her back on the road to recovery. Her accident taught her to view the world with a heightened awareness of the here and now. Too often, says Eileen, the past is dragged up or the future contemplated, but all anyone really has is today.

Eileen remembers the response she once got from a friend when she had voiced a number of complaints: "Eileen, always remember that problems are there to be overcome." In farming, says Eileen, problems are never far away, and often arrive in droves. This further underlines the importance of a listening ear, of someone for farm families to turn to when life has dealt them a blow.

Eileen's spontaneity has paid dividends. "One morning, I got a phone call from a colleague who told me that there were a couple of places left on the IFA tour to Eastern Europe, and did I want to go. Without any clear idea of how I would pay for the trip, I decided on the spot to take up the offer. I never regretted it, and the trip was a truly memorable experience."

Farming has been central to Eileen's life. It is what defines her. This has not prevented Eileen from cultivating interests outside farming, such as study and travel. She was delighted to be awarded the IFA/FBD scholarship, which took her to the Baltic states, which otherwise she might never have seen.

"I am sure that my sense of justice stems from my negative experiences in primary school. I believe that had I not become a farmer, I would have been an activist in whatever field I found myself."

Elizabeth Tilson

Elizabeth has been there for me from the start. Her outsize personality, lovely warm smile, wonderful sense of place, and her great attachment to country life, have been infectious. She also is very open to change, and it was great to observe how she responded to the Australian trip.

One of Elizabeth Tilson's first memories is clinging to her mother's leg on the first day of primary school in September 1968. "A neighbour's boy was doing the same to his mother, while both women were engaged in earnest conversation with the teacher. Despite this unhappy start to my educational career, I survived my first day, and enjoyed school enough to stay right through to my Leaving Certificate."

Elizabeth was born to a farming background. She was the oldest of four children (she has two sisters and a brother), and was the first grandchild on her father's side of the family. At a very early age, Elizabeth suffered the trauma of losing her father. He died suddenly just a few months after the birth of her youngest sister.

Elizabeth's mother was left to look after four young children, and to carry on looking after the family farm. "I grew up knowing that all of us were very precious in our mother's life. My mother did everything she possibly could for the children, and I feel that me and my siblings were given more from her than some children get from both parents. We went to the scouts and the guides, and we were encouraged to pursue music, swimming and other extra-curricular activities."

After she left school, Elizabeth did a secretarial course. In the middle of her studies, she successfully passed the entrance examinations and interviews for nursing school. She secured a place to study nursing in England, and left for her studies in October 1981. One can imagine the consternation of Elizabeth's mother when her daughter arrived back home just two weeks later.

"I came home because I hated everything about the nursing school. But it was more than that. I was also seriously concerned about my mother being on her own, now that my brother and sisters were away at boarding school. Of course, my sudden return created a new concern for my mother - what was she going to tell the neighbours and the family?"

During the summer before she went to England, Elizabeth's family had become the first people in the area to build a slatted shed. On the first morning after Elizabeth's return from her short-lived London experience, her mother's fears were confirmed. An uncle who was visiting them found Elizabeth grapeing (forking) silage into the cattle in the slatted shed. His words to her that morning left an indelible impression on her: "You're an eejit for coming back, you will end up grapeing shit for the rest of your life."

Elizabeth was accepted to study nursing in Belfast starting the following April. In the meantime, she heard of an office position in the local Dairy Cooperative. She applied for the job, was interviewed, and secured the position. She liked the job so much that when April came around, she decided that she would stay put rather than going off to take up her nursing job in Belfast.

Elizabeth describes how she met and married her husband Mervyn. "A few months after I was born, some new neighbours moved into the area. They held a housewarming party, and since there was no baby-sitter available, my parents took me along. Also attending the housewarming party was a little fellow of four years old who took an extraordinary interest in the tot in the cot. He had no brothers or sisters of his own, and decided that I would make the perfect playmate."

Back to the summer after Elizabeth returned from her short-lived England experience. "Thanks to my mother's insistence, I had become a keen swimmer, and I also loved sailing. I joined the local canoe club, and on St Stephen's Night I attended a canoe club disco in Virginia. Here I met a guy who had come to the disco with my friend. The three of us went off to Longford to attend another dance. The guy took a sudden intense interest in the canoe club, and after a couple of weeks he plucked up the courage to ask me out."

Their first date did not go as planned. In fact, it never materialised at all. It was an evening of very bad snow, and Elizabeth and her work colleagues had to walk home on top of cars and trucks, following the route according to the tops of the trees that were the only thing emerging from the blanket of snow. Naturally, they had to fix another date, but the delay does not seem to have done any harm, because the two of them were soon a regular item.

"Imagine our surprise when we discovered that our meeting in the Virginia disco had not been our first encounter. That's right, he was the very same boy who 20 years earlier had taken a fancy to me in my cot at the housewarming party."

The couple got engaged at Christmas, and decided to marry the following August. In the meantime, Elizabeth started going to evening classes in crafts and needlework, for which she soon discovered she had a real aptitude. She progressed to doing Carickmacross Lace, which is how she came to make her own wedding veil.

Elizabeth gave up her job a week before the wedding because she would no longer be earning enough to justify the travel to work. The young couple moved into the farmhouse vacated by her parents-in-law, and settled into married life, even though there were some who commented that they were more like brother and sister than husband and wife. In 1987, the birth of their son was greeted with much happiness, especially because he was the first grandchild in the family on both sides.

"My life was turned upside down by a tussle with my health two years after I gave birth. I became very ill, and spent two weeks in hospital. The tests were inconclusive, even though it should have been clear from looking at me that this was a case of jaundice. An old woman who occupied the next bed in the ward said to me, 'Daughter dear, get someone to go to two brothers who live beside me, they have the cure for jaundice.'"

Elizabeth's husband went to the brothers who handed him a cure to take to Elizabeth. The cure was meant to be administered three times, on a Monday, Thursday and the following Monday. But when the doctors did their rounds in the hospital on the Tuesday, just one day after the first treatment, they were astonished to see that all signs of the jaundice had disappeared. Elizabeth was released from hospital the next day.

After a few weeks at home, Elizabeth was feeling at a loose end, so her husband suggested that she seeks seasonal work in the nearby meat factory. She could not quite believe that was offered a job on the spot, and she turned up for work at 7am the next day. At first, she was sent to work in the canteen because one of the regular canteen workers was always taking days off. However, a supervisor came and said she was not needed in the canteen, so Elizabeth was sent to the boning hall, where she surprised herself by enjoying the work.

It was the period before Christmas, a time when traditionally people were laid off because this was the end of the seasonal work. One of the foremen asked Elizabeth if she would be prepared to stay on. She agreed, even though this did not go down too well with some of the other women, who firmly believed in the last in, first out principle, regardless of ability or performance.

"Several months later, I was called into the office of the production manager. 'What have I done wrong?' I asked myself in trepidation. I needn't have worried. The reason for the summons was that the boss wanted to offer me the position of Store Manager that was soon to become vacant. The management believed that I was capable of taking over the procurement of products and to keep the factory running smoothly. I agreed on the spot, and that very same day, I started working in the store to learn the ropes. Once I took over, I discovered that I loved being in charge."

Elizabeth had to cover all aspects of the procurement process, and to make sure that there was fuel in the tanks to run the factory, and to keep the kill floor and the boning hall equipment working. She had to source packing material for the beef, and to buy the food for the canteen. She had to provide First-Aid, and to accompany anyone to hospital in the event of an injury. It was a high-pressure job, but she thrived on it.

When Elizabeth thought she was ready for a change, she gave up the job in order to devote more time to her husband and son, and to the farm. She had managed to put away enough money from her earnings at the meat plant to invest in some Simmental cows, which she gradually became involved in showing. She worked hard teaching the animals to halt, walk after, and walk with her, her efforts being rewarded when she started winning awards. She became regional secretary of the Simmental Cattle Breeding Association and remained in the post for seven years.

"My first involvement with the IFA began in 1996. At first, I handled membership in the county, travelling to the different branches with the branch officers. I would visit the local farmers, and listen to them commenting on what IFA should and shouldn't do. I took all this in my stride, and became more and more comfortable with the inner workings of the IFA. I eventually became Assistant County Secretary, and then County Secretary. "

One of the highlights of Elizabeth's IFA career was the January 2000 beef blockade, when farmers manned picket lines outside major meat plants around the country. They were protesting against low prices and in favour of the removal of the meat inspection levy. Elizabeth used her administrative skills to keep the rota going, making certain that there always enough people manning the different shifts. Working in tandem with other county officers, she made sure there was always enough soup and sandwiches to go round. Her canteen supply skills that she had learned in the meat factory canteen came in useful.

"The factory blockade had its own special challenge for me. I suddenly found myself on the other side of the fence from my former employers, the management of the meat plant. But in fact, this worked to my advantage. I became a sort of unofficial liaison between the factory and the blockaders. The management would phone me at the factory gates or even at home if they wanted lorries to leave. I would speak to the farmers at the gate to ascertain their decision. If the farmers said no, I relayed the decision back to the management, and they accepted this."

In 2001, Elizabeth got her first taste of national IFA politics when she canvassed on behalf of one of the candidates for the IFA presidency. Running the county campaign taught Elizabeth a great deal about what people will say about each other in the heat of political debate. But the experience was enjoyable and she gained a lot of true friends.

In September 2001, Elizabeth decided to finally do what she had originally set out to do twenty years earlier, and she enrolled in university. "In November 2004, I graduated with a Diploma in Rural Development, and I am on course to complete my degree in Rural Development in 2006."

The experience of having to depopulate her herd in 2002 because of Brucellosis made a lasting impression on Elizabeth. There was actually only one animal in the herd with a high reading, but because they were in a zero tolerance area, the entire herd had to be depopulated. For Elizabeth and her husband, this was a terrible loss, as they had successfully built up a top quality replacement stock for both the dairy and the suckler herds.

"For three weeks, the animals were constantly in and out of the crush for testing, tagging and valuing. On the wet and dreary morning they were due to be taken away, some sixth sense gripped the cows, and as if they knew it was the end, they refused to come in. For me, there was added sadness and poignancy in seeing the line of trucks on the road just down from the farm, parked as far as the eye could see, waiting to take the animals away for premature slaughter."

Still shaken by what had happened to her herd, Elizabeth decided to put herself forward for nomination as the County Animal Health Committee Representative at the IFA County AGM. She believed that her experience with herd depopulation gave her a special advantage. Although there were other candidates, Elizabeth's passionate desire for the post won the day, and she became County Representative at national level.

"When I turned up for the first meeting in the Farm Centre in Dublin, the first order of the day was to elect members of the Animal Health Management Committee. Many of the people there were new, and I did not know who to vote for. In fact, I did not even put my own name forward. Imagine my surprise when I was voted on to the management committee for the northern half of the country."

Elizabeth has thoroughly enjoyed her time with the IFA. "I was particular happy when the IFA appointed their first-ever Equality Officer. I was proud to be part of the steering group that hosted the highly successful International Women's Day event at Croke Park in 2003. And, of course, I thoroughly enjoyed the super 'Women in Agriculture' trip to Australia in 2004. It certainly whetted my appetite for another trip down under."

At a personal level, Elizabeth says she has gained a great deal from the personal development workshops and training sessions for IFA women. "In particular, I value the tips I learned on how to present myself confidently, and how to run meetings more effectively."

Bernie Murphy

Bernie is a strong and determined woman whose bravery and insights came to my attention early on in my IFA career. I have always been able to pick up the phone and know that she would offer some new angle to help me out. She is a brilliant listener, full of encouragement, and in some ways even a bit of a rogue. I'd say that Bernie's sense of fun means that she is usually the last person to leave a gathering.

"Growing up in County Wexford in the fifties, I firmly believed that Wexford was the only county that really counted in Ireland. So strong was this conviction that I almost pitied people from other counties. It was therefore something of a shock, when I went to study in Dublin at the age of 18, to encounter people from far-away counties. It was even more of a shock to discover that people from Cavan were quite normal!"

Bernie's mother was from a wealthy farm background, and her father worked as a farmhand for a family of Protestant farmers. Because Bernie's next sister up was seven years her senior, she was pretty lonely during her childhood. Her memories of those early years are of playing alone, and feeling more like an only child, than the youngest of four. She rarely interacted with her siblings, and the only person in her own age group was a cousin who was a couple of years older.

Bernie's mother had a good head, and thanks to her practical ways, the family was virtually self-sufficient. Most of their food came from what Bernie's mother grew in their vegetable patch, and their protein needs were met by the hens that were raised in a coop in the garden. Even the family's bedlinen was made out of recycled flour sacks.

"I was very attached to the people who owned the farm where my father worked. Most of the brothers and sisters in the family were unmarried, and they were like family for me. I called them all Uncle and Aunt, and even as a child, I sensed that I was privileged to know such gentle, good people who lived simply and never flaunted their wealth. On an evening, they were often to be found sitting in our home playing cards with my parents."

These uncles and aunts loved and pampered Bernie, and she felt closer to them than to her blood relations. They allowed Bernie to build a baby house in their hayloft, and she loved playing with their dogs. There was only one thing that disturbed Bernie's idyllic view of her uncles and aunts. She had been authoritatively informed that these lovely people would not get into heaven because they were not Catholics, and Bernie spent many a night crying herself to sleep worrying about this terrifying thought.

Bernie was never a tall person, and as a child walking to school, she would easily tire after about the first mile. Just as she would start crying from exhaustion at the one-mile mark, she and her sister would always be joined on the walk to school by a strapping lass of a neighbour. "For a couple of years, she carried me the remaining three miles to school, and the two of us still have a good laugh at this memory."

The young Bernie did not particularly like school. "Things bothered me. It struck me quite forcibly at school that land gives a person status. I noticed that the children whose parents had land enjoyed greater prestige than children like me who did not own land. I also remember being puzzled by the fact that so many priests attended a funeral mass for a rich person who had died, and that a much smaller number of priests would attend the funeral of a poor parishioner."

Secondary school - a boarding school run by nuns - did not prove much happier for Bernie than junior school. She objected to the upstairs-downstairs inequality between the teaching nuns and the domestic nuns. The teaching nuns, who came from wealthier families and brought a good dowry to the convent, lived in single cells upstairs. The domestic nuns, who came from poorer families, lived in dormitories down in the basement, next to a perpetually noisy boiler. "I felt a sense of injustice. Why was there discrimination between rich and poor, even within the Church?"

Probably because of her small frame, Bernie experienced a lot of bullying at boarding school, and her first few years there were not happy ones. She felt self conscious and intimidated by girls that she perceived to be from wealthier homes. Eventually, things started to improve. She has happy memories of going to mass, especially at Christmas and Easter, walking around with veils, and singing hymns in Latin without understanding a word. It felt mysterious.

"The bullying only stopped when I proved my prowess on the hockey field, and when I established a reputation for being funny. For a time at school, I went through a very religious phase, and even thought of becoming a nun. But I gave up this idea when I discovered boys and started regularly getting into mischief. Once, me and a group of friends "borrowed" bicycles, and rode to the beach where we smoked. Someone snitched on us, and as we were sneaking back into school, a nun was waiting for us at the school gate. This escapade, plus other occasions when I sneaked out to meet boys, almost got me expelled."

Bernie found the rules and regulations at boarding school needlessly oppressive. She objected to the fact that one even needed permission to wash one's hair in the middle of the week. Bernie never really took to her schoolwork, and in retrospect she feels that she would have benefited from home tutoring if there had been such a thing in those days. She believes that her personality would have been more suited to self-directed learning. She was disappointed – but maybe not overly surprised – when she failed her Leaving Certificate by one subject. She was especially disappointed for her parents, who had worked so hard to allow her to get an education.

Before Bernie even sat her Certificate, she had passed the entrance examination for a nursing course in a Dublin Nursing Home for people with emotional disabilities. Just prior to commencing her nursing studies, Bernie went to her first big dance – and was overwhelmed by the big band, the noise and the craic. She boarded at the nursing home, which was run by nuns – not too different to her boarding school.

"I found living in Dublin an eye opener. I met and made friends with girls from around the country – including County Cavan. I combined nursing studies with hands-on treatment of the long-term patients, and my particular interest was occupational therapy. If I'm not mistaken, the patients used to pack plastic cutlery for Aer Lingus flights. A highly gifted lay occupational therapist at the nursing home inspired me to want to get involved in this field."

During her first year of studies, one of Bernie's Protestant "aunties" developed breast cancer. Bernie accompanied her to Dublin, and took her to Grafton Street. This may well have been the aunt's second-ever visit to this famous street. After having tea and scones at the late-lamented Bewley's Café, Bernie bought the aunt a bunch of flowers from the street vendors. The aunt was so touched by this spontaneous gesture that she burst into tears.

Bernie's plans for a career in occupational therapy were thwarted when she became ill with chronic back pain. She had to give up her studies, and was angry and disgruntled with God for forcing this change of plans. Bernie stayed on in Dublin for a few more months, sharing a flat with some friends, and going out to work in a shop. She became part of swinging Dublin, drinking too much and attending wild parties.

"But I also slipped into depression. I returned to my parents' home, where I was greatly incapacitated by back pain. I helped a little around the house, and tried to get myself into better physical and mental shape. While working in a local office, I heard of an opening for someone to run a community pre-school. I attended childcare courses, and together with a voluntary helper, I was appointed to run the pre-school. Over the next couple of years, I worked happily with the youngsters, and engaged in a sometimes hectic social life."

Throughout this whole period, Bernie did not really know who she was or what she would like to do. Today, she feels that she and her generation were severely held back by Church-induced guilt about their sexuality and personality. One evening, Bernie met Jim at a dance.

"He fell for me immediately, and seemed to see in me a whole lot that I myself could not acknowledge at the time. It took me a while longer to make up my mind about marrying Jim, because I wasn't yet certain whether I wanted to give up my wild side. But I have never regretted my decision to succumb to the charms of the man who, according to my friends, managed to tame me!"

Jim was the eldest son of a farming couple. He owned part of the family farm, and was in line to inherit. Bernie and Jim got married when Bernie was 23, and built a home on the farm. For Bernie, moving to Jim's farm felt a bit like being under the microscope, but she soon became adept at keeping the peace among the different members of the family. Bernie continued to run the pre-school for a few months after the wedding, but from the birth of her first child, she fulfilled everyone else's expectations and became a homemaker. She ran the home, helped run the farm, raised her children, cared for her parents and parents-in-law, and fed the farmhands.

Bernie has always taken an active interest in running and building up the farm, and husband Jim has always regarded her as a full and equal partner. He has always been ready and eager to share everything with her – family, farm, and IFA decisions. Bernie regards herself as the organisational brains behind her husband, and for his part, Jim has always wanted to make Bernie feel included.

"Although I continued to enjoy working on the farm and raising my family, I would let my hair down at the weekends. I loved recapturing some of the daring of my schooldays, and I used to enjoy the occasional thrill of after-hours drinking in the local. I grew close to my mother-in-law, even though we used to argue over child rearing practices. I now feel that I probably kept a lot bottled up, because things were not discussed openly like they are today."

Over the years, Bernie and Jim struggled to keep their asset rich, money poor farm going, and kept their heads above the water. All revenue went towards growing the farm, rather than towards home improvements or home extensions. Bernie says that farming is an all-consuming effort, and you have to give it your all. She sometimes feels that she might have given a little too much attention to the farm, and too little to her home and children.

Jim was always involved in the IFA and other farming organisations. Although Bernie was also interested in farming issues, there were precious few outlets available to women in those days. But Bernie is not too quick to let women off the hook. " I believe that it is up to women to go after what they want. I don't accept the conventional wisdom that IFA women are put down by the IFA men. I think that women often put themselves down. They don't aspire to achieve enough, can sometimes be their own worst enemies, and too often put other women down.

In the early eighties, Bernie started becoming more actively involved in IFA affairs. She remembers participating in demonstrations against the Irish government's introduction of the Milk Super Levy in 1984. She was a great help to Jim when he became chair of the local IFA, and used to phone the farmers to get them to attend meetings.

It was during this period that Bernie herself joined the Farm Family committee. She remembers turning up at one of her first meetings and shocking some of the other women there. "I said something about we women should be fighting for our own rights, and that there should be a tax allowance for the work women do on the farm. This was long before the IFA Equality Initiative, and long before demands such as these became mainstream."

Bernie has never been afraid to express her own independent views, even if they have not been particularly popular. On many issues, she was way before her time, and some people thought her ideas were foolhardy at best, and quite mad at worst. But today, when many of these ideas have now become part of official policies, Bernie feels vindicated. She believes that too many farm women are giving their energy to the Farm Family Committee, and not enough to broader and more mainstream farming sectors.

Bernie took over the chair of her local IFA branch from husband Jim. "The key to success in IFA is to get yourself noticed. Women have to fight harder to get their ideas out there, but I welcome the challenge. I firmly believe that women should be more proactive, and less afraid of their own shadow. I also believe that many traditional role definitions are outdated. I am not keen on the automatic stereotyping of men's and women's roles. Equality is an absolute right. It is not in my nature to be overawed by status and power, and I am much more impressed by values such as sincerity and integrity."

Since childhood, some of Bernie's fondest memories have involved the outdoors. One spring day, not too long ago, she went out with her dog. The cattle were lying in the grass, the trees were swaying in the breeze, and she suddenly spotted a hare and a fox, eying each other. The hare was preening itself, and nonchalantly walked off. The fox seemed quite nonplussed, and went off in the opposite direction. Magic moments and idyllic scenes like these remind Bernie of how lucky she is to be alive.

Despite the sickness and bereavement that have sometimes dogged her, Bernie is a determined survivor. "I am on a continuous quest for self-empowerment, and would love nothing more than to be known as a woman who did it my way."

Mary Flynn

Ever since I first met Mary when the Australian women came to visit in 2003, I have looked on her as my right-hand person. Mary took on the whole organisation of the Australians' itinerary in Ireland whilst I worked on arranging the conference. It was also Mary who was instrumental in organising the subsequent trip to Australia in 2004.

I love the fact that Mary is always so full of positive vim. Her family and farm mean so much to her, and she is a great believer in walking as a means of getting in touch with herself. Every time we meet, Mary gives me such a fantastic hug, holding onto me for dear life – surely a sign of her wonderful open personality.

Like a lot of unassuming women, Mary Flynn is reluctant to acknowledge that her life has been anything except ordinary. She does not believe that she has achieved anything of real significance in her life, and she still does not understand why her story should be included in this volume. At most, Mary is prepared to admit that in her own small way, she might have made quite a considerable journey.

That journey began far, far from any contact with the farming world. Mary was the eldest of four children, and her parents both worked in the hotel and catering business. Mary's father, Teddy McGivney, worked as a chef, and Mary's early childhood was spent in Tramore, County Waterford, where he was Head Chef in the Grand Hotel. Mary's mother Rita ran a guesthouse, also in Tramore.

"When my father got the opportunity to become Head Chef in the newly-built Tower Hotel in Waterford City, he accepted the post, and we made the move to this considerably larger metropolis. My childhood was spent in a happy and loving home. I remember that when we weren't at school, we all used to spend much of our leisure time outdoors, playing football and tennis on the streets."

Mary and her younger sister Gaye were sent to Scoil Mhuire boarding school in Carrick-on-Suir. All subjects were taught in Irish at that time. "To this day, I still find that I am more comfortable with certain expressions in Irish than in English. This was an era when boarding school was a much more rigid place than today. Boarders rarely went home between the beginning and end of term, and my sister and I were only allowed to leave the school for a home visit on the first Sunday of every month. How we lived for those precious days off."

Because Mary's social life centred so strongly round her boarding school, it was inevitable that her school friends became like a second family. This pattern has persisted, and to this day Mary counts many of the girls from Scoil Mhuire as her friends.

"My mother got it into her head that I was too shy and quiet, and that something had to be done about it. I suppose I must agree that my mother's diagnosis was correct. So when I returned home after leaving school, she decided to do something proactive, and encouraged me to volunteer my services to the Waterford Credit Union. At the time, credit unions were purely voluntary organisations, and they were constantly on the lookout for willing helpers."

Mary's mother's strategy proved successful. Mary's encounters with the members of the public gradually helped her overcome her shyness. Within a few months, as Mary says now, "I was able to put chat on anyone I met there."

Mary continued her volunteer activities at the credit union every Friday night for several more years. She loved every minute of it, and is thankful for her mother's foresight in pushing her to become involved.

Mary's plan after leaving boarding school was to enrol to do a business course in the Waterford Institute of Technology. She was accepted, and commenced her studies. But midway through the course, she applied to and was accepted by the Royal Liver Assurance company. So before she sat her final exams, Mary left college to take up her new job.

"This was at a time in Ireland's history when jobs were hard to come by. I was worried that such an opportunity might not come up again when I left college, so I decided that full time employment now was better than possible unemployment further down the road. I took quickly to the work, and within six months, I was promoted to the post of Principal Clerk."

It was during the first year of her work at Royal Liver that Mary met the love of her life, Gerard Flynn. He had everything going for him, except one important thing - he was a farmer. "I had dated farmers before, and my previous experiences had convinced me that there was not much hope in such a relationship. This did not prevent me falling in love with Gerard. Farmer or not, he was the one for me. We became inseparable, and I soon overcame any lingering anxieties regarding his chosen profession. We were married in 1976."

Until the 1970's, an established feature of life in Ireland was that in many sectors of the economy, women had to give up their jobs when they married. If Mary had been a couple of years older, this would doubtlessly have been her fate as well. However, it was her good fortune that the marriage bar was lifted just a few months before she got married. For the first time in Ireland, it was now acceptable for young women to be able to hold on to their jobs after they entered the state of holy matrimony.

So Mary continued to work for Royal Liver in Waterford, and drove the twenty miles there and back every day. Remarkably, for someone who had been plagued earlier in her life with shyness, Mary now started taking an active part in trade union affairs in her company. Although victory of a kind was achieved with the lifting of the marriage bar, a bar of a different kind – the equal pay bar - had most certainly not yet been lifted.

"I was shocked and dismayed to discover that the men in my office were being paid more than the women for doing exactly the same work. I fought hard to secure equal pay for myself and the other women, but with little real success. By the time I left in 1979, equal pay for men and women was still a distant and unattainable dream. How far Ireland has come since then."

The reason that Mary left Royal Liver was that she gave birth to her first daughter in 1979. This effectively marked the end of her full-time working career.

Mary now found herself spending all her time on the farm. She no longer spent most of her waking hours travelling to, and working in, Waterford City. She decided that now that she was more involved in the day-to-day farming, she wanted to become more knowledgeable about farm matters. The city girl with no farming background enrolled on an ACOT (now Teagasc) course. "The more I studied, the greater I developed a real interest in agriculture, and in what was happening on the farm."

Gerard and Mary had started with a calf to beef enterprise. They then went into tillage, before getting into cows in 1983. Whenever Mary tries to recall her days in the milking parlour, she can always see a child's pram in the scene. As her children were growing up, they were always around as their parents got on with the job of farming. It was therefore no great surprise that all of the children later developed a great love of farming.

"The ACOT course I attended gave the participants, all of us women, a sense of greater empowerment. So some of us got together and organised ourselves to form the first IFA Farm Family Committee in Waterford City. For a while, the committee thrived, but some of the initial enthusiasm was already waning by the time I took over the chair. Attendance gradually fell away even further. At the time, I thought it was my fault. I was sure that my old lack of confidence that my mother had identified many years earlier, was the reason for the committee's demise. I convinced myself that it was my own inability to motivate others that was the cause of the smaller attendance."

A few years later, when the opportunity came in 1992 to re-establish the committee, Mary was determined to make amends. This time, she was sure she could succeed in galvanising the other members, and she accepted the role of secretary of the IFA Farm Family Committee in Co. Waterford.

In September of the same year, something happened that was to have a major impact on Mary's view of life. She now had four children, including two who were still very young. On this particular Tuesday, Gerard had taken the older two children to the National Ploughing Championships. It was agreed that Mary would stay at home to look after the milking.

"Because the younger two girls were so young, I had asked our neighbour Maurice Boland if he would come over that evening to give me a hand with the milking. He did so, and it was when the cows were in the collecting yard that I was suddenly attacked from behind by a bull. To this day, I have no recollection of the incident. All I remember is waking up in the ambulance, wondering what I was doing there. If Maurice had not happened to be around at the time to call the ambulance, I doubt if I had lived to tell my tale."

Mary believes that the goring incident changed her priorities and affected her attitudes to a wide range of issues. From that day, she started regarding every new day as a bonus. She looked hard at her usual way of handling things, and made a conscious decision to stop fussing so much about trivial things.

Her new enthusiasm for life also translated itself into her work as secretary of the Farm Family Committee. "I started a campaign to push the notion of self-improvement training to the top of the agenda. As someone who was born into a generation where anyone who was too outspoken was called a "notice box," I was only too aware of the barriers that women faced. Girls of my generation had been brought up to be nice, quiet and passive young ladies. Without any serious alternatives to choose from, we were quite happy to fit in to the straightjacket that society placed us in."

But all Mary's experience in life since childhood had taught her that this upbringing had not done women any favours. Her memories of the inequality of pay at her employers made her doubly convinced that she had a duty to ensure that her fellow members of the Farm Family Committee would have the opportunity to receive training.

Over the years, Mary and her colleagues have been responsible for organising short courses in confidence building and self worth. Mary has availed herself of each and every opportunity to further learn and improve.

With the approach of both the new Millennium and her fortieth birthday, Mary once again decided that it was time to take stock of her situation. She felt she was ready to make a re-assessment of her priorities, and she identified several areas where there were still important things to be achieved.

"I always attached great importance to education. I used to wonder if I had made the right decision when I interrupted my business studies in order to take a safe job with Royal Liver. When I discovered that University College Cork was offering a two-year part-time Diploma in Women's Studies in Clonmel, this was the trigger that pushed me into enrolling as a student."

It did not take very long for Mary to find out that a third level course is a far cry from being in school. It wasn't that she felt out of her depth in higher education, on the contrary, she absolutely loved the freedom of being able to express herself fully. She loved the fact that she was now expected to have, and to express, her own opinions, and to voice her own views in class and in her essays.

"At school, I had always had difficulty in filling up one side of one page when handed an essay assignment. Suddenly, in college, I found myself able to write 2,000-word assignments without a problem. What's more, I made the even more surprising discovery that I could achieve consistently excellent grades for my written work. I managed to sail effortlessly through my course, and my graduation in October 2002 stands out as one of the best and proudest days of my life."

Mary was one of many farmwomen who greatly applauded the appointment of an IFA Equality Officer in 2001. The existence of a full-time officer charged with promoting the cause of women within the IFA dovetailed perfectly with Mary's own ideas of continuing education for farmwomen. "Mary Carroll's appointment as Equality Officer coincided with my taking over the chair once again of the

Farm Family Committee. With the support and encouragement of Mary, I saw that when I put my mind to something, there was nothing I could not achieve."

For years, Mary had fantasised about going on a trip to Australia. This idea remained in the realm of an unattainable dream, with the long distance and the prohibitive cost just two factors that seemed impossible to overcome. Little did Mary know that it would be through IFA that she would achieve her goal. As soon as the IFA Women's Network announced that they were planning a study trip of Australia, Mary knew that come what may, she would be part of the trip. She managed to find the fare, and together with 32 other Irish 'Women In Agriculture', she embarked on what she describes as the trip of a lifetime.

"With another major milestone – my 50th birthday – fast approaching, I'm busy setting new goals and ambitions for the years to come. The girl who left her business course after school to go out and get a job, has become a knowledge-hungry woman. I am actively contemplating the notion of going on and completing my MA in Women's Studies."

Maura Horgan

I had barely been a week in my new job as IFA Equality officer when I first heard the lovely bright Cork accent of Maura Horgan coming over the phone lines, full of positive encouragement. She had read about my appointment, and took the trouble to ring up and wish me well. It was the start of a friendship that has been very enriching.

Maura's knowledge of pedigree cattle and cattle breeding is amazing! She is truly an expert, and could easily become an international advisor on the subject. When Maura decides to do something, you can be sure she will deliver. She works exceptionally hard, she never stops, yet through it all, she still manages to look absolutely stunning.

Recently, an Irish family achieved fame with a mention in the Guinness Book of Records - because all eight children in the family never missed a single day's school in their entire educational career.

Maura makes no claim to be part of that particular family, but when she was growing up, her own family certainly came close to emulating that feat. Maura is the fifth child in the family. She has two older brothers, two older sisters, and a younger brother and sister. As a child, Maura and her siblings used to walk the four kilometres to and from school every day, trudging to school in every kind of weather – rain or shine, sleet or snow. And because of the start and finish times, there was never any question of transportation to and from school.

Farming has held a fascination for Maura since she was born. Her family lived on a farm in Glanmire in County Cork, where her father was employed as a farm labourer. When Maura was five months old, her parents bought their own farm in Dromduve, Ballylickey.

"I remember spending many an hour playing in the yard at milking time. It was only natural that I lent a hand on the farm, but from an early age, I developed an abiding interest in the animals. This fascination with cows, pigs and sheep has stayed with me throughout my life."

Maura's carefree childhood came to an abrupt halt during her first year at secondary school, Ard Scoil Phobal in Bantry. Tragedy struck the family when Maura's father was diagnosed with cancer, and within a few months he was gone.

Maura's older brothers and sisters had already left home. This left Maura's mother to fend for herself, rear three young children and run the farm.

Before Maura's father died, he had made it clear that he wished his oldest son to inherit the farm. But although the farm did indeed pass on to Maura's brother after the untimely death of their father, he showed no real interest in becoming actively involved in its running. He already enjoyed his work putting up sheds on other farms, so the day-to-day running of the family farm fell on the shoulders of Maura and her mother. Maura rose to the challenge like a champion.

"When I was just twelve and a half, I would be up each morning at 6am, joining my mother in milking the cows. Then I would go off and put in a full day's work at school. As soon as I arrived back home at the end of the school day, I would grab a bite to eat, and off I went again, looking after all the things that needed attention on the farm. It was very hard work, but I discovered early on that I thrived on hard work."

For a while, everything ran smoothly. Despite the adversity, the family managed to survive. Maura somehow kept abreast of her school studies, and her schoolwork did not suffer unduly. Her mother was delighted with the help she was getting from Maura.

There was great family celebration when Maura's oldest brother announced that he was getting married. The wedding was especially memorable for Maura. "It was here, at the age of fourteen, that I first clapped eyes on Eamonn, who soon became my boyfriend."

After the brother's wedding, Maura's mother found a job, and gave up her work on the farm. Although the brother took a little more interest in the farm, it was Maura who was left shouldering the entire responsibility for the running of the farm. At the time, this did not strike Maura as unusual. And she was so innocent that she never even thought of raising the issue of being paid for her efforts.

Through no fault of her own, Maura became embroiled in family politics. Her new sister-in-law seemed to take exception to the splendid job that Maura was making of the farm. Not that the sister-in-law had wanted to be personally involved in the day-to-day running of the farm. It was just implied that Maura might one day get greedy and put in a claim of her own for the farm.

"Such a thought had never once entered my head. I could not understand how anyone could confuse my keen sense of responsibility with scheming manipulation. This problem was only really solved when I left home to marry. I had been dating Eamonn for several years, and although I was still young, we decided that we wanted to spend the rest of our lives together. Despite the disapproval of my family, Eamonn and I married in 1983, shortly after I sat my Leaving Certificate."

At seventeen and a half years of age, Maura was now a married woman. True, she had no degree, and no real idea of what her life was going to be like, but she felt she was the happiest girl alive. The early days of married life were no bed of roses, but the fighting spirit that she had displayed while single-handedly running the family farm, now came in very useful.

Eamonn had a piece of land with a new bungalow. After selling several cattle to pay for the wedding, Eamonn was left with just a few cows. When they first married, the young couple had precious little money, and Eamonn was only earning about £3 an hour for work as a machine driver. Maura became pregnant with their first child, and did not work. They had so little money that they could not even afford to have electricity in their bungalow.

"Our luck started to change after the birth of our baby boy on Christmas Day. Eamonn found a better-paid job, and there was now a little more money around. Eamonn started to buy calves, which at the time cost between £20 and £30. Our financial situation improved considerably when a neighbour offered to rent about 30 acres on our farm for a period of five years. This gave us a new degree of stability, and Eamonn was able to buy a few more calves with the proceeds. By this time, I was pregnant again."

Eamonn and Maura applied for a grant to wire the remaining 40 acres, and they were sanctioned a grant of £7,000 which allowed them to get their place up and running. Although Eamonn was worried that he would lose money from his paid job if he took time off to do the wiring, there was little choice. Under the grant terms, if the wiring was not finished within a certain period, the grant would have to be returned. So Eamonn took off one day a week. He and Maura applied themselves to the task at hand, and did as much wiring as they could, often working in the dead of winter when the days were short, cold and wet.

Shortly before they completed the wiring, their second son was born in February 1985. Money was still tight, but by the early summer, the job was done and they managed to beat the deadline. The work was seamlessly approved by the Inspector, and the grant went straight into their bank account. "I remember that we never actually saw the money, but I also remember the relief of knowing that at least the pressure had eased. We no longer faced any danger that the bank would throw us out of our home."

That summer, in order to feed whatever few cattle there might be in the coming winter, Maura and Eamonn brought in a contractor to make a pit of silage. The calves that Eamonn had bought were doing nicely, and as winter approached, they were put near the forestry for shelter. That winter, Maura learned to drive the tractor for the first time. This enabled her to take over the feeding of the animals while Eamonn continued in his paid job.

The following spring, they put an Angus bull out with the heifers that they had wintered. The heifers were due to calve in February 1986, and throughout the summer and autumn months, it was Maura who looked after them and checked that everything was in order.

Once again, the heifers were wintered in the forestry, but this time it was harder going, especially when the heifers started to calve out at night. This had to be done under lamps, but luckily none of the cows strayed too far. The calving was successful, and Maura was delighted that they would now have some weanlings to sell at the end of the year. Cattle prices were very good at the time, and it was possible to make over £500 per head, which in those days was considered a lot of money.

"In January 1987, at the age of twenty-one, I gave birth to our third son. That summer, we decided to expand into sheep. Eamonn did not know too much about sheep, but I was more than happy to teach him whatever I knew from my own time on the family farm. We bought lots of lambs that autumn. We also made the disastrous decision to buy some old ewes, but we did sell our first lambs in the summer of 1988. The lambs were heavy, and fetched a healthy £47 each."

The second winter of their sheep enterprise proved more difficult than the first. Maura and Eamonn decided to lamb the ewes outside, but it started to snow just as lambing began. If the lambs are picked up immediately after birth, they die. Luckily, there was an old house on the farmyard. The house had actually been sold, but since the people were gone, Maura and Eamonn hastily converted it into a shelter for the ewes during the lambing. This innovative solution saved the day, and very few lambs were lost.

Maura could not face the prospect of another winter like that again, so in 1990, they borrowed £5,000 and put up a sheep shed. They now had enough sheep and cows to cover repayments, and Eamonn continued to bring money in from his job. Maura and he did most of the work on the sheep shed themselves, including putting in the floors. Although she was pregnant with their fourth child, Maura used to join Eamonn at night and work on the shed after the boys had gone to sleep.

The shed was completed in late December 1990, just days before Maura gave birth to Rebecca on New Year's Day. "I was so proud to have produced a daughter after three boys. But I wasn't allowed the luxury of celebrating. In quick succession, we were hit by a series of misfortunes. On the very day I came home from hospital, there was a message for Eamonn that his mother had suffered a stroke. She died a few days later. On the day of her funeral, Eamonn and I were feeding the cattle before leaving for the church. I was trying to hurry Eamonn up, because some of the funeral arrangements still had to be attended to. Suddenly, one of my brothers arrived to tell me that our mother had also had a stroke that morning."

Somehow, Maura got through the trauma of these upheavals. All the experience and training that she had gained as a twelve year old when she practically ran the farm alone after her father died, now came in very useful. This resilient woman settled down to take care of her enlarged parcel of responsibilities: her new baby, her three sons, her mother, Eamonn's step father, and of course the farm.

That winter, Maura and Eamonn put up a small lean-to off the sheep shed for the cows. Plans to put in cubicles didn't work out, because there was no place to put all the dung and slurry. By the time they put up their first slatted unit in mid-1992, they had a Whitehead bull and about 24 cows. Since they knew that continental weanlings were always dearer at the end of the year, they bought a Charolais bull that produced prize-winning cattle at weanling sales.

The slatted unit was finally completed, and by that winter, Maura had the satisfaction of seeing her sheep and cattle indoors. Bale silage made life much easier for Maura to manage, and the sheep used to get hay. Whatever had to be done to the sheep and cattle, whether it was dipping or dosing, Maura and Eamonn would always do it together. At first, the younger children would go to a friend who lived a short distance away. Then, as the two older boys grew up more, they too started to help in the work.

Throughout all this period, Eamonn continued being employed outside the farm. Maura continued to do all the feeding and to look after the general day-to-day running of the farm. She also helped Eamonn's stepfather on his farm, and she and Eamonn always hid the fact that Eamonn was working off the farm. The stepfather always wanted Eamonn to be around in case he needed him for something. So whenever he rang for Eamonn, it was Maura who had to drop everything and attend to him. It was a challenge to balance this with running the farm and raising four children, but Maura was good at meeting challenges.

In the spring of 1994, Maura was pregnant again. This time, she could not go near the sheep, so Eamonn had to give up work for the lambing. He was absolutely determined that every ewe would be sold. It was tough having to do everything himself, but Maura's health and the health of their unborn child were paramount. One day in early July, Maura helped Eamonn draw in bales of silage. The very next day, she gave birth to their second girl, and their fifth child.

"Unfortunately, the pattern of celebration and tragedy was repeated. Eamonn's stepfather died just a few weeks after the baby was born. Eamonn inherited his farm, and it was clear that I couldn't manage our family as well as both farms on my own. That's when we decided that it was finally time for Eamonn to give up the security of his paid job."

The newly inherited farm was very run down, and needed a lot of work. Maura and Eamonn set about wiring the farm all round the boundaries, and they also did some reseeding. When this was completed, they had to put up a slatted unit. Maura had misgivings about borrowing too much, as they had just about repaid all their outstanding loans. But there was not much choice, so they borrowed £25,000 for the fattening unit and the yard. Eamonn now devoted himself full-time to the farms. They put up a large hayshed at home for storing bales of straw and hay.

"I had always had a hankering to get into the breeding of pedigree cattle. That's where I believed that money was to be made. With some of the pressure off now that Eamonn was running the farms, I found time to complete a Farm Management course. The ewes were sold out as the cow numbers increased, and I devoted myself more and more to the breeding aspect of our farm business. I knew exactly what I wanted - to produce prize-winning cattle on a consistent basis."

After all these years, Maura was farming the way she had always wanted to farm. Cow numbers were growing, and Maura had the satisfaction of fulfilling her dream. She did indeed start proudly producing prize-winning cattle. To house the pedigrees, Maura and Eamonn put up another slatted unit, and as usual they did all the work themselves. At the last count, Maura had 65 suckling cows, many of them purebreds, a mixture of Belgian Blues and Limousin Purebreds. The rest are cross breeds.

Eventually, things were running so efficiently that Eamonn was able to return to outside work on a full-time basis. Today, once again, Maura is managing the farm single-handed. The boys are grown and all have jobs. When something special needs to be done on the farm, the boys all turn up and help Eamonn out on a Saturday. The girls are still in school.

Anyone reading this story of the young girl who started running a farm on her own at the age of twelve will not be surprised that this extraordinary level of hard work and determination was also recognised by others. The judges in the 2001 All Ireland Farm Woman contest were so impressed that they decided to award Maura first prize. For Maura's family, friends and colleagues, here at long last was recognition for her work and her ceaseless efforts.

"When people ask me what keeps me going, I always reply that it is my love of farming. When I look around, it saddens me to see so many farmers being pushed out of farming. I also feel that farmers get a bad press, and tend to be misunderstood by the general public."

Maura was never someone to sit on her laurels. She has recently taken over the running of another farm some 35 miles away. She milks 50 cows morning and evening. 21 years after she got married, Maura still works about 16 hours a day. She insists on staying up with all her cows when they are calving.

"This is my way of protecting everything I have worked for. This is the only way I know to do things – by putting my heart and soul into everything I do."

Rosemary Kennedy

Rosemary is an energy pack. I have always found her fiercely independent and determined to go it alone as much as she can. Rosemary had to take over the farm when she became widowed, and she has built it up with gusto and determination. She is a real symbol of strength, a woman who can serve as an example for us all.

On what was undoubtedly one of the highpoints of her life, Rosemary stepped up to accept her award from Minister of Agriculture, Joe Walsh, following the announcement that she had been chosen as the 2000 Bank of Ireland National Farm Woman of the Year. As she made her way to the podium, few people in the audience could have guessed the tragic circumstances in which Rosemary had suddenly found herself nine years earlier.

That January weekend in 1991 was going to be an occasion of fun for the whole family. On the Friday evening, the Kennedy family – husband Richie, along with sons Keith aged 14 and Richard aged 9, made their way to Carrick-on-Suir, County Tipperary. Here they attended an awards ceremony where both boys received medals for under 14 and under 12 titles. Some weeks earlier, Rathgormack GAA football club had won both juvenile county titles.

Rosemary remained at home to look after 11-month-old Fiona. The prize ceremony ran a little late for the boys, and it was a tired but cheerful crew that arrived back at the farm. As they slept that night, a violent storm raged outside. The following morning, Rosemary allowed herself a rare sleep-in while Fiona and the boys slept soundly.

As usual, Richie was up early to feed the cows, cattle and sheep. Once he had completed his own tasks, he travelled a short distance by tractor to help his brother Johnny feed his animals.

As a result of the storm, telephone wires lining the home farm avenue had been blown down, and now lay strewn across the laneway. Richie had to weave his tractor between the wires on his way to his brother's farmyard. When he heard that his mother and sister intended to take a trip into town, Richie quickly retraced his steps to cut the offending wire and move it out of the way. Richie did not notice that the wire had connected with the overhead ESB power lines, and he was instantly electrocuted.

These were the thoughts that passed through Rosemary's head as she made her way to receive her award.

Rosemary was born on a small family farm in Powerstown, a rural parish on the outskirts of Clonmel, County Tipperary. She met Richie in the summer of 1974, and a few months later, Rosemary, then 19, lost both her parents within a few days of each other. They had both been suffering from heart disease, and Rosemary had been caring for them for some time. Rosemary and Richie married less than a year later, and settled on Richie's 145-acre family farm in Clondonnell, Rathgormack. This farm, at the foot of the scenic Comeragh mountains in County Waterford on the border with South Tipperary, was just 12 miles from Rosemary's family home.

Rosemary and Richie both quickly set about the task of farming their mixed enterprise of Suffolk sheep, beef and tillage. They also built themselves a house. "Over the years, economic rationalisation and changes in EU legislation persuaded us to purchase Friesian heifer calves in the spring of 1983, as we planned to build a dairy herd. However, we changed this plan when the introduction of the milk quotas was announced. We purchased a Hereford bull, and we gradually phased out the tillage enterprise. During this period of change, three children arrived to keep things lively."

It was this farm enterprise that the suddenly widowed Rosemary now had to run. Still in her mid-thirties and with a young family, she found herself at the helm of a farm for which she had no special training and precious little idea of what she should do.

Rosemary did not want to throw in the towel. "I decided immediately to remain farming. Quitting was never an option. I really didn't want to leave the farm that we had both worked so hard on together, and I never considered moving the children away from their home. It was their home, and I felt I had to carry on for the sake of Richie's memory and for my children's future."

After Richie's death, friends and relatives assumed that Rosemary would keep Keith at home to help run the farm, but she never contemplated that for a moment. Instead, she insisted that Keith first complete his schooling, which he did, before going on to earn an honours degree in Agriculture from the University of Wales at Aberystwyth. He is currently teaching Agriculture with Teagasc, but still regularly returns home to be of great practical and emotional assistance to Rosemary. Richard also went on to further education, and after earning an honours degree in Physical Education and Sports Science, he took up a teaching position in Cornwall.

By any yardstick, those first few years after Richie died were very tough on Rosemary. She was determined to master the complexities of running the farm single-handed. To achieve this, she knew she needed to bring herself up to speed very rapidly.

Did she ever have doubts about her abilities? "Frequently." Had she made the right decision? "No one else could answer these questions. Just me." Plagued by doubts, Rosemary often turned to the livestock for solace. "I would often go up to the farm and sob my way through the work that had to be done. I didn't want my children to see me so upset, they had enough sorrow themselves to deal with, without also worrying about me."

Rosemary embarked on a sharp learning curve. Until the accident, she had only ever assisted at a calving. In those first weeks, she would ring Johnny or a neighbour if she realised that a cow was having a difficult calving. But one day, no one was available, so Rosemary managed alone. From then on, she never looked back, although she is still not afraid to seek assistance when it is needed.

One of her first business decisions was to sell their flock of sheep as they were due to start lambing. This was shortly after the funeral, and because the cows were also due to begin calving, Rosemary felt she could not manage both without a lot of help. Instead of keeping the sheep, she began increasing the size of the suckler herd. Over the next three years, she purchased Simmental heifers from neighbouring farms.

In 1994, Rosemary drew up a farm plan with her Teagasc advisor Paddy O'Brien. She built slatted sheds and extended and adapted other sheds, replacing the silage slab and aprons over the next couple of years. But no sooner had she managed to get the farm back on an even keel when misfortune struck, in the form of a bad outbreak of Tuberculosis. A few weeks earlier, Rosemary had moved her cattle into their new sheds.

Following a herd test, 19 of the 24-month-old steers tested positive, and 6 further cases emerged in two subsequent tests. All of the cattle had been earmarked for selling the following spring, but the TB outbreak meant that the herd now had to be locked up until the following autumn. This put her finances under severe strain, but true to form, she refused to buckle under.

"It was the day before Christmas Eve, and even though I saw everything crumbling around me, I knew I had to put my feelings aside and prepare for Christmas. My children needed and deserved this. I think the two boys were more devastated by the herd test results than I was."

Adhering to her farm plan. Rosemary began a major reseeding programme that included the reseeding of 64 acres over a seven-year period, as well as drainage, land reclamation, paddocking and providing new water supplies to the farm.

Over the years, the suckler herd profile has changed, both in size and in breed type. Rosemary initially changed from Friesan to Simmental and some Hereford, but in 1996 she purchased a Limousin bull, and a second one in 1999. She changed from multi-suckling to a single suckler herd, and changed from purchasing replacements from dairy herds to producing all her own replacements. In 2002, she bought a Simmental bull for replacements, and in 2005 she introduced a new Limousin bull to continue the genetic improvement of her herd. Her hopes for a closed herd that would be protected from disease were dashed when the herd contracted BVD (Bovine Viral Diarrohea).

Rosemary is someone who seems to thrive in the face of adversity. In addition to running the farm and raising her family, this feisty woman began a dizzying programme of study and self-improvement. She completed the 20-month Advanced Certificate in Drystock Management at Kildalton Agricultural College, Piltown, Co Kilkenny, under the auspices of Teagasc. She was the only woman on the course, which included visits to drystock operations in Belgium, France and Spain. She obtained a Diploma in Community Education and is studying for a Certificate in Archaeology from UCC, and always keeps abreast of the latest developments in farm affairs through talks, seminars and discussion groups.

It was Rosemary's interest in farm affairs that led her to want to actively participate in farming politics. She was elected branch chair of Rathgormack IFA, and subsequently became Waterford IFA Rural Development chair. In 2005, Rosemary was elected to the Executive of County Waterford Community Forum, and she is on the board of management of various schools, community centres and clubs. She holds very definite views on bureaucratic inefficiencies. "I have gone on record highlighting the hardships involved in applying for the new waste management and dairy hygiene schemes."

In 2003, Rosemary was a member of the FBD-backed IFA international fact-finding tour of the USA. She visited one of the world's biggest agricultural shows, the World Dairy Expo in Madison, Wisconsin, which is attended by tens of thousands of visitors. Here she saw some of the latest technological developments, and she also attended North America's top dairy cattle breeding shows.

It was the grit and determination displayed by Rosemary over the years that proved major factors influencing the decision of the judging panel to award her the 1999 Regional Farm Woman of the Year. The following year, Rosemary hit the jackpot when she won the national title. In particular, Rosemary impressed the panel with her acute business acumen, her ambition, and her foresight in developing her skills through constant education and training.

In October 2004, Rosemary joined 31 colleagues on the 'Women in Agriculture' trip to Australia, where they attended 'The Tall Poppies' conference as part of the 2004 World Rural Women's Day. "Among the farmwomen that I met on the trip were members of the Australian delegation to the Croke Park conference a year earlier, some of whom had stayed with me in Rathgormack. The whole trip was an experience to be savoured. I found that despite the very different landscape and scale of farming, Australian and Irish farmers have similar concerns. I was particularly impressed with the Australian way of tackling succession planning, with all members of the family sitting down with a facilitator and an accountant to discuss together the sensitive issues involved."

Rosemary joined some of her Irish colleagues in Tasmania where they visited self-sufficient farm families. Some of Rosemary's hosts were part of a Tasmanian delegation that visited Ireland in 2005. She has also travelled to China as part of a Teagasc study trip.

Rosemary is very conscious of the fact that without the willingness of her family to fill the void while she is away on her travels, these trips could never take place.

"I am especially grateful to my brother-in-law Johnny and his family. Without them, I doubt if I'd still be farming. Special thanks too to my daughter Fiona, a fourth year school student in Scoil Mhuire, Carrick-on-Suir, my friend and constant companion over the years. I look forward to her spreading her wings and fulfilling her dreams. I also want to thank my sons Keith and Richard for their unswerving love and loyalty."

The day that changed Rosemary's life for ever is a fitting tribute to finish her story.

"Some years ago sitting in my car at my outside farm going to check on my cows and calves. I was listening to Gerry Ryan's show on the radio and something jolted my senses. I picked up a piece of paper and luckily I had a pen and in a couple of minutes I had written this poem about the moment I was told of Richie's death".

Gone she said, gone where I cannot know
My love's gone, the father of my children
Blown away, to a place I can only dream of
Hollywood scenes filter through my frozen brain
Gone! I could not fathom the reason
The reality of which I shied, cried away from
My sons uncomprehending faces stare at me waiting for truth
My sleeping girl child slumbers
Blissful in her ignorance of life
My infant cries in solidarity, I scream inwards
My peaceful morn evaporated in black reality

Grainne Dwyer

It was natural that I would meet Grainne early on in my IFA career, as she had written a Nuffield scholarship report on women in agriculture. I kept coming across her name, and Grainne has been a trusted advisor ever since. I love her intellect and her helpful insights. She has strong opinions and is determined to change things. It is Grainne's brand of determination and strategic skills that will help move things on for women in agriculture.

In 1943, Lord Nuffield, famous as the owner of Morris Motors, established The Nuffield Foundation. Under the Nuffield Farming Scholarships Awards that were first announced in 1947, Nuffield Scholars were encouraged to develop their leadership skills in farming. The scholars would be funded for a year to allow them to study farming practices and techniques employed anywhere in the world.

In 1998, Ireland joined the scheme, and the first Irish people were awarded Nuffield scholarships. In 2000, Grainne Dwyer became the first woman ever to be awarded an Irish Nuffield Farming Scholarship.

"I grew up on a small farm in Kildare. My father was an auctioneer, and he ran a mixed sheep, cattle and tillage farm on a part-time basis with the help of the family."

After completing her Leaving Certificate, Grainne got a job in a firm of solicitors in Newbridge, and in general she felt that her life was going nowhere. Although she quite enjoyed the work, it presented her with little challenge.

"My next job, working in the accounts department in the Law Society in Dublin, proved to be something of a revelation. I travelled up every day from Newbridge, and I simply loved everything about the job. I enjoyed the companionship, and I felt both nurtured and encouraged by management. Around this time, solicitors decided that the prohibitive cost of professional indemnity demanded that they create their own scheme. At least 1,000 solicitors had to join the Mutual Defence Scheme in order for it to work. Once this target was reached, the scheme went into operation. I was delighted to be appointed to administer the professional indemnity scheme."

Grainne took to the job like a fish to water. She discovered that she had considerable administrative skills, and she received great support from her boss, the then Director General of the Law Society, who had a particular gift for spotting potential.

During this same period, Grainne started attending the local branch of Macra Na Feirme, the young farmers association. This was the perfect place for someone with her positive enthusiasm, and between 1983 and 1985, she served as County Secretary and County Chairperson. In 1986, Grainne was chosen to represent Ireland at a European young farmers rally. This entitled her to join a team of four Irish farm people and attend a conference in Germany that brought together young farmers to discuss common farming interests whilst having fun.

"My boss encouraged me to participate. He saw this as a feather in the cap of the Law Society, and he liked the idea that a member of his staff was representing Ireland abroad. In Schleswig Holstein, where the conference was taking place, I met the other members of the Irish team – including my future husband Jim."

They say that love conquers all, but it was not easy decision for Grainne to move down the country to become "the Dairy Farmer's Wife." Grainne had just progressed from being County Secretary to County Chair of Macra, but she barely had time to tackle the problem of dwindling membership before she had to leave the post. It was also difficult to give up her highly enjoyable and high profile job at the Law Society.

"I found moving to Jim's farm in Borris in Ossory in 1988 a huge wrench. I had to give up my salary, my status and my public positions, and get used to having no regular job, no salary and the new status of 'blow-in'. I made a conscious decision about farming. I accepted that this was going to be my livelihood for the rest of my life, so it was time I learned more about farming. I wanted to contribute to the farm business, but I didn't know the basics."

Easier said than done. As Grainne looked in vain for opportunities where a woman can learn about farming, she began to resent the fact that there was nothing available that suited people in her situation. Doing a course in a college was not an option, because it would involve having to be away from home, which she didn't want.

So in the end, it was Jim who provided her with her farm education. He patiently taught her everything. Of course, this was also in his best interests, but Grainne still claims that she received the best education possible.

Grainne's first five years of married life were marred by a series of tragedies. Both her parents died, she gave birth to a stillborn child, and she and Jim were involved in making a family settlement. But finally, in 1993, they got full ownership of the farm, and Grainne gave birth to a healthy boy, Jonathan. Jim was now able to introduce some of the radical changes he had been planning, and this was an exciting time for both of them.

Jim and Grainne originally thought that they would stay with grass-based farming but it needed further refining. They began to learn more about how things were being done overseas, and they were eager to be among the first to introduce new ideas into Ireland. Jim had always done things differently. He had always been prepared to buck the trend, and Grainne's personality was similarly inclined.

"We started looking seriously at the New Zealand model, which involved more grass and longer grazing time. This low-cost production model was cheaper than housing the cows. We still had access then to milk quotas, and because the farm was growing, it was no longer viable for us to have the cows exclusively outside. In 1998, we decided that if we were to make important decisions, we needed to do some serious research. So we went to New Zealand to check things out. We spent three weeks there, studying and refining the New Zealand model. We discovered how New Zealand dairy farmers coped when the weather was poor and the cows were damaging paddocks, by putting their cows in a stand-off (out-wintering) pad. We returned from our trip very excited, ready to implement the new ideas we had learned."

Grainne had long believed in the need for farmers to master advanced business skills. So when the opportunity to do a business course came along, both Jim and Grainne joined a group of 15 farmers for an intensive training schedule that covered assessing the farm business, evaluating other business models, buying into business, equity and investment. The course proved an important turning point for Grainne and Jim, and they were now able to bring a much more businesslike approach to their farm business.

"In 1999, we built the first stand-off pad in Ireland. This generated huge interest from other farmers, many of whom came to see the pad for themselves. But this was essentially Jim's pet project. I decided that I needed something of my own. After the business course, I was primed to go, and it was in this state of mind that I decided to apply for an Irish Nuffield Farming Scholarship."

As there are only two scholarships given in the whole of Ireland each year, Grainne knew that the competition would be tough. She also knew that a woman had never before been awarded a Nuffield scholarship in Ireland. But she believed she had a good chance. She knew they were looking for people who are ready for something new, ready for something different, and ready to think outside the box.

"I travelled to London, with the other two final candidates, with some confidence. My real challenge was to convince a panel of eight men that I was a suitable candidate to conduct a study entitled: 'Women: Access to Agriculture', which explored how women are neglected in farming. The panel was farsighted enough to award the scholarship to an Irish woman for the very first time."

Under the terms of the scholarship, Grainne had to devote one year to her project, including twelve weeks abroad. First, she had to identify where to conduct her research, and to set up contacts. In this she was helped and inspired by an extraordinary woman, sociologist Dr Patricia O'Hara, author of 'Partners in Production'. In her book, Patricia had revealed that the paucity of statistics regarding women farmers in Ireland, and indeed in all EU states, is such that more information is known about the animals and crops on family farms than about the women.

"I set out on my Nuffield year with gusto. During my week in Norway, I discovered that the eldest child inherits, whether son or daughter. Next, I visited France, and on this trip I was accompanied by Jim. It was while I was attending a conference in Holland on gender equality that I met Sally Shorthall, from the Department of Sociology and Social Policy in Queens University, Belfast who was also a great help and inspiration."

In January 2000, Grainne, Jim and young Jonathan travelled to New Zealand, where Grainne conducted her research while Jim continued his study of local farming practices. Jim and Jonathan then returned to Ireland at the end of January for the calving, while Grainne continued on to Australia.

"I spent five weeks meeting some of the most amazing women and it was an honour and my privilege to meet with them. I also met up with other Nuffield scholars, current and past. I found them warm and welcoming, and totally on my wavelength. These women were fully focused on the women's agenda, and I felt myself rebelling against the passive acceptance by too many Irish farmwoman of their situation. I met women who had made a difference, women who were making things happen, especially in dairying. I was shocked to discover that the milk cheque is issued in both names – husband and wife."

Grainne discovered women were recognised as equals, they operated in Australia's dairy farming sector in the open market, with no artificial props. The open market is a very different system and farming within it can be challenging, so both partners need to be focused. But the women were regarded as farmers in their own right. She met women who knew as much, and sometimes more than their husbands, about the business. She met people involved in government supported equality roles, such as the Equality Officer for Rural Affairs, and the head of the Rural Women's Unit.

"All my research substantiated the fact that the work performed by women on the farm in Ireland is not recognised and is almost invisible. I realised that Ireland was way behind more enlightened countries like Australia, where the government established a Rural Women's Network almost two decades ago. Back in 1992, the Australian government acknowledged the scope of the problem when it issued a report on the contribution of women to agriculture, entitled 'The Invisible Farmer'. I returned home, and starting writing my report. I wrote that in contrast to the Irish experience, women farmers in Norway, France, Australia and New Zealand receive full recognition as farmers from their respective governments, semi-state agencies, and rural communities. In Ireland, women farmers enjoy no such recognition. They are known as farmers' wives, rather than farmers in their own right."

For Grainne, the Nuffield period was a journey of self-discovery and self-growth. In the period since she completed her report, Grainne has realised that not all Irish farmwomen share her passion about the need to address women's issues in farming. She has also discovered that many women who do reach prominence in farm associations, actively seek to keep other women down. This goes against all that Grainne stands for. As a believer in a bottom-up rather than a top-down approach, Grainne sometimes wonders if Irish farmwomen will ever get the plot.

In 2001, Grainne became secretary of the Irish Grasslands Association. She loves working with a professional non-political agribusiness organisation, and is busy organising four conferences a year: beef, sheep, and two dairy conferences. This job she does in conjunction with running the farming business with Jim.

"Nuffield has left a lasting impact on me. Through Nuffield, I came into contact with a lot of forward-thinking people around the world who share my positive attitude, and my dedication to progressive thinking. I continue to meet up with my Nuffield peers at international Nuffield conferences, and my trips abroad give me the adrenalin to help me overcome some of the pockets of negativity that still persist here at home. My goal is to help make farmwomen more comfortable in a man's world. I am against preferential treatment for anyone, including women. I don't think women need special conditions. They should be truly equal in thought and deed."

Clearly, Grainne's quest for self-discovery and empowerment is far from over.